The Colder the Better

Books by David Wilson

The Colder the Better *(1980)*

In Search of Penicillin *(1976)*

The New Archaeology *(1975)*

Body and Antibody: A Report on the New Immunology *(1972)*

DAVID WILSON

The Colder the Better

New York 1980 ATHENEUM

Published in Great Britain under the title SUPERCOLD

Library of Congress Cataloging in Publication Data

Wilson, David, 1927-
The colder the better.

Originally published under title: Supercold, an introduction to low temperature technology.
Bibliography: p.
Includes index.
1. Low temperature engineering. I. Title.
TP482.W54 1980 621.5'9 79-637903
ISBN 0-689-10873-7

Manufactured by Fairfield Graphics, Fairfield, Pennsylvania
First American Edition

Foreword

It is perhaps a little too defensive, but this book must begin with two apologies, both aimed to pacify those scientists who may read it.

The first is for the fact that all temperatures in the book are not carefully labelled with the correct C., K. or F. Such labelling is, of course, essential when scientists are communicating with each other professionally, for it would be stupid to give a measurement of temperature without stating the system in which it is made. But for the general public the multitude of temperature scales, with their complications of pluses, minuses and differing zeros, is confusing. I have striven for clarity rather than conformity with convention.

The second apology is due because the names of the many scientists who have helped me are not given here at the start of the book. Instead I have listed them at the end among the references. References are supposed to indicate the authorities on which statements in the text are made. For a journalist writing about scientific matters those scientists who have provided information, explanation, guidance and criticism *are* the authorities, though in this case any mistakes must be blamed upon the author rather than upon them. Dr T. Faber must be mentioned with special gratitude for the quantity and quality of his help.

Hampton Hill
August 1978

Contents

The Colder the Better

1 Prometheus

The myth of Prometheus, who stole fire from the gods and carried it to men hidden in a stalk of fennel, expresses a profound truth which we recognize as part of our cultural history and heritage: fire is the symbol of man's leaping imagination, his intellectual creativity. But science, too, sees truth in the Prometheus myth. The conquest and control of fire seems to be the first great step that distinguishes man from the other animals. To science, fire symbolizes energy; the use of fire is the key to releasing the resources of our planet.

The scientific interpretation of the myth of Prometheus heightens our appreciation of this piece of ancient wisdom just as much as the cultural interpretation. The ability to control energy, and to direct its application, enables us to control our environment. Fire can be used to make things cold as well as to make them hot. And our western, urban, technological civilization depends on making things cold just as much as making them hot.

For hundreds of centuries it appeared that technological progress went hand in hand with producing and controlling higher temperatures. In the last one hundred and fifty years we have discovered that equal or greater advantages can be obtained by forcing matter to ever lower temperatures. And while we appear to be approaching the limits of our ability to control the highest temperatures that we can visualize producing, the biggest expansion of opportunities for exploiting low temperatures seems still to lie ahead of us.

In our western culture we know the Prometheus myth from the writings of Homer and Hesiod, at least five centuries before the start of our era. But the appreciation of the importance of controlling fire lies far deeper in the folk-memory of many human cultures than the earliest writings. Similar myths exist in most of the Indo-European languages, and in the Vedic tradition. 'Fire-bringers', birds or men, are part of the mythology of Australian aborigines and the Norse legends of the Isle of Man. In many traditions fire was brought by a demigod who was responsible also for most of the basic advances in technology.

The Greeks, however, did not regard Prometheus as a technology god—technology was rather the province of Hephaestus. Prometheus was the man who deceived the gods. He deceived them not only over fire, but also at a meeting of men and gods held to decide which portions of sacrificed animals should be allotted to men and which to gods. Here Prometheus hid the meat of the animal underneath the offal, and the bones underneath the fat. Zeus chose the rich-looking, fatty pile, and never forgave Prometheus when he discovered there were only bones underneath. Prometheus, according to some versions of the myth, was chained forever to a rock and had his liver torn out regularly by birds of prey. Whether this punishment was given for stealing fire or for deceiving Zeus is not entirely clear.

The scientific, or evolutionary meaning of the myth may well have been elucidated by the most recent discoveries of archaeologists and anthropologists working in East Africa. It has been popular to call man 'The Toolmaker', and to regard the first distinction between man and the remainder of the animals as being the ability to use objects external to himself in order to achieve his aims. Essentially this ability is conceptual as well as technological, for toolmaking implies the capability of abstract reasoning towards a planned goal, at least two steps away. However, it has recently been shown that chimpanzees can prepare a number of tools from twigs and leaves and use them for obtaining food and cleaning themselves after feeding.

It is also reasonably certain from the findings of the Leakeys and others in East Africa that the ability to make stone tools was not confined to *Homo sapiens*, but was shared by creatures such as *Homo habilis* and probably by the hominids such as *Australopithecus* who may have been our ancestors or our cousins. On the other hand, the earliest archaeological evidence from Africa that seems to imply the deliberate and controlled use of fire apparently coincides with the evidence for the emergence of the first truly human cultures of *Homo sapiens*. With discoveries still being made at a great rate in this exciting field, the exclusive relation of fire to true humans may well have to be abandoned or the dates changed.

It remains true, however, that the controlled use of fire continues to distinguish man from the other animals even now. Animals can, and do, make and use tools. Animals certainly communicate with each other. But no animal, except man, uses fire, keeps fire, or deliberately makes fire.

Evidence of the first use and control of fire can never be obtained. The existence of the Promethean myth is as near as we are ever likely to get to understanding this vital step in human evolution. Archaeology can provide nothing (as yet) except evidence that fire was used in ways which must have been deliberate. It is guessed that fire was snatched from the fringes of some natural catastrophe, a tree stricken by lightning or a bush fire. It is assumed that the ability to make fire deliberately by some process of rubbing came many generations later. In the meanwhile the keeping of fire was a task of enormous importance which, it is assumed, played a role of fundamental importance in the formation of social relationships and religious rituals and practices.

It is known that man used fire to improve tools, which may have been weapons or digging sticks, by hardening the points. We extrapolate from our own experience in guessing that man used fire to protect himself from wild animals and to warm himself. We have no idea when or how man discovered that fire could be used for cooking his food and improving its palatability and digestibility. Our experience with the use of cold in food preparation over the past one hundred and fifty years would suggest that the use of fire for cooking was a far more important factor in technological development than most of the academic authorities have thought.

Despite uncertainties of detail, it is clear that the use and control of fire by man represented enormous steps forward in the development of technology and the ability to control his immediate environment, which in turn allowed him to occupy almost every corner of the land surface of the earth. But even more important is that aspect of fire which we describe as energy. Fire is a redistribution of the energy that is locked up in the biosphere, the thin life-containing shell around our earth in which we exist.

The energy available to primitive man in the biosphere came from the sun, powering the motion of the seas and the atmosphere, powering all the chemical reactions and rearrangements which we call life. (There is also a vast quantity of energy locked in the very structure of the atoms that make up all matter. This energy is derived from virtually unknown cosmological processes and history. We are just beginning to use and control it in our nuclear age, but it plays little part in the subject matter of this book.) Fire is one of the many chemical reactions, or rearrangements of atoms and molecules of matter, which release energy. The use of fire is the use of this released energy to cause other chemical reactions to take place; that is, to cause fresh rearrangements of matter which can only occur when extra energy is supplied. Some of these fresh rearrangements can be of types which rarely, if ever, occur naturally. The energy released by fire which can cause these fresh rearrangements we feel, and describe, as heat.

After the primitive uses of fire for protection, warmth, cooking and tool-hardening, the most important application was for making pottery. This development is generally held to have occurred at about the same time as that momentous step in human history called variously the Neolithic Revolution or the Agricultural Revolution—in other words, the development of farming and a settled way of life in place of the nomadic life of hunting and food-gathering which had preceded it. It now seems probable that there were many different, separate and possibly independent 'Neolithic Revolutions' in different parts of the world, but the earliest took place in the Middle East some time around 10,000 BC. The earliest pottery is found in the Middle East at various sites and various dates beginning somewhere about 8,000 BC. It is possibly pure coincidence that the invention of pottery came at roughly the same time as the invention of agriculture and the first development of settled life in villages and early cities. But the association of pottery with the rise of civilization has been accentuated by the simple fact that pottery, breakable but virtually indestructible, easily transportable but an essential feature of domestic life, simple but capable of carrying the mark of an individual manufacturer, has been found to be the most durable and the most easily identifiable of all human artefacts and hence the single most important type of evidence available to the archaeologist.

Chemical analysis of early pottery makes it clear that man comparatively soon began to increase the temperature of his kilns above that which could be obtained by simply burning any of the fuels available to him. He must have forced extra air into the kiln; at first, presumably, by blowing through pipes. Later there must have come primitive bellows made from animal skins. We guess this by observing the processes still in use in poorly-developed societies. But the pottery produced by higher temperatures was undoubtedly better—'finer' in both senses of the word.

From this time onwards, for the six or seven thousand years to our own time, man has developed his technological achievements and abilities by steadily achieving higher temperatures and controlling greater amounts of energy.

After pottery came metallurgy. There is fierce argument among the archaeologists as to whether metallurgy began in the same area of the Middle East that saw the start of the Neolithic Revolution, or whether it began in the Danubian area of the Balkans. There are also arguments in favour of an entirely independent start for metallurgy in China. In the Americas, where the great civilizations of Central America had apparently invented pottery independently for themselves, there was no metallurgy till the Europeans arrived. It is suggested that the first metalwork was simply the beating into shape of naturally occurring metallic copper outcrops. There may, on the other hand, have been accidental discoveries of the melting of metal in pottery kilns.

However this may be, the sequence of 'The Three Ages' that we have all learnt (Stone Age, Bronze Age, Iron Age) is a statement of the ever-increasing temperatures attained by technology. Bronze smelting requires higher temperatures than dealing with malleable natural copper metal. Iron-making requires higher temperatures still. This 'progress' in technology is usually described in terms of invasions and tribal movements, for metal weapons were stronger than those of stone and wood, and iron weapons were stronger than those of bronze. But it is equally true that iron ploughs enabled the farmer of Celtic and Roman Europe to cultivate heavy clay soils that were not amenable to the tools of his predecessors. Better axes likewise enabled thicker woods to be cleared before the plough.

Though the technological achievements of the Middle Ages have probably been underestimated, it is reasonable to point out that the materials available to the people of Renaissance Europe were no different from those available to the Romans. From about the fourteenth century onwards, however, small quantities of steel began to be produced for weapon blades. Steel was almost certainly discovered by accident in the process of reheating iron to higher temperatures while 'tempering' the edges of swords and daggers. It was not until 1856, however, that large quantities of steel could be produced when Bessemer perfected his process.

Before that, the Industrial Revolution had started. Although this advent of our modern world was a vastly complicated development, in the long run the Industrial Revolution can be seen as the large-scale application of energy, through steampower, to human activities. It was the discovery and control of steampower that enabled the Victorians to spread factories over the world. It was steam that powered the machines with which the Victorians performed their feats of engineering. Our modern mass-transport systems of railways and international shipping were created by steampower.

Steam at higher temperatures still drives almost every electric power station in the world, and the development of the Electric Age in which we live is no more than the control of even larger quantities of energy, and the ability to distribute that energy in a clean and convenient form to millions upon millions of outlets in homes, factories, offices and farms all over the world.

The technological progress of the last half of the twentieth century is a repetition of the same story over and over again—more efficient engines, providing or concentrating greater quantities of energy, have continuously been sought and found by operating at ever higher temperatures. The only limitation to this progress has been the problem of finding materials which can operate at the temperatures we wish to attain. The power attainable from jet engines is a typical case in point—the power has gone up directly as new metal alloys have been developed to work at higher and higher temperatures. Our houses are better illuminated than those of our ancestors because we have found a way of producing and maintaining very high temperatures, and therefore very bright light, in tungsten wires.

The first nuclear electric power-stations, built as recently as the 1960s, gave out their heat at temperatures less than 300°C. Nuclear stations of the 1970s operate at 600°C. We are now experimenting with fast-breeder reactors which need liquid metals to cool them. We look to the future and hope to control nuclear fusion (the power of the hydrogen bomb). The temperature of the fusion reaction is measured in tens of millions of degrees. We know we can find no material to operate at this temperature, so we plan to contain the reaction within rings or bottles of enormously powerful magnetic fields. It appears that we are approaching the end of the trail. The drive for new materials to control higher temperatures is probably ending in the latest developments of rocket engines. The fusion power stations if they ever come will use energy (magnetic) to contain energy production.

The last two hundred years have seen an enormous acceleration and intensification of the process begun by the first control of fire. Now it seems that the end of that process or the limits to it are in sight. The process has gone from the snatching of a burning branch off a stricken tree to the flash, brighter than the sun, of the hydrogen bomb. The control of fire and the use of heat have been but one of man's lines of technological advance, but whereas the discovery of the wheel, or the domestication of animals and plants, has only led to the modification of the materials found available on earth, the use of fire and heat has allowed man to make materials never found in nature.

While the last two centuries have seen the final rush into the technologies of applying heat, they have also been marked by the use of ever higher temperatures in these new technologies. In the same period we have seen the emergence and even more rapid development of the opposite set of technologies, those dependent on the creation of cold, the removal of heat. Just as the 'hot' technologies seem always to have been improved and expanded by the ability to reach higher and higher temperatures, so the 'cold' technologies seem to achieve greater and greater effects as the temperature is lowered. This is the theme of this book—'the colder the better'.

Low-temperature science and technology began only after the Industrial Revolution had started. There is virtually nothing worth recording before 1800, and very little except some 'pure science' before 1850. The first large-scale use of low temperature technology which had any serious political or social effect started in 1873 and was not affecting our society in any important way a hundred years ago. Low-temperature science and technology is, then, a very recent factor in human life and history.

The man who is generally accepted as the originator of modern experimental science died as a result of an experiment in low-temperature effects. Francis Bacon, eventually Lord Verulam and Viscount St Albans, was a typical courtier of the first Queen Elizabeth. Under James I he attained the higher rank of Lord Chancellor but was dismissed and disgraced for bribery. In the last five years of his life he achieved far greater work by enunciating the basic principles of experimental science, and this thinking led to the foundation some fifty years later of the Royal Society of London. Besides writing about science, Bacon practised it. Travelling one cold and snowy day in March 1626 near Highgate, he suddenly decided to 'experiment' and see whether snow would delay the putrefaction of flesh. He stopped his carriage and purchased a fowl on the spot. He stuffed the creature with snow with his own hands, and in the process was attacked by a sudden chill. Although he was nursed at the nearby house of Lord Arundel, the cold and chill developed into what we now call bronchitis, and he died on 9 April, 1626.

Fascination with the preservative power of ice and cold continued for almost two centuries without any real understanding of the processes involved. Nor was any useful experimental work performed to provide explanations. Fifty years after Bacon's time Pepys, the London diarist, was fascinated by reports from Baltic merchants that by the use of ice they could keep chickens throughout the winter without decay. A hundred years later, in 1799, the Russians reported finding the frozen bodies of mammoths in Siberia. It was considered one of the wonders of the world, and the fact that dogs, at least, could eat the thawed flesh of the mammoth added to the interest.

At roughly the same time some scientists, or 'natural philosophers' as they were then regarded, noted that cold sufficient to freeze water was produced when volatile liquids such as ether were allowed to evaporate while in contact with water. (Evaporation is the transformation of liquid into gas; it occurs at different temperatures and pressures according to the chemical nature of the liquid: it is most commonly seen when water boils, which occurs at 100°C at normal atmospheric pressure.) The first attempts to make ice artificially used the evaporation phenomenon, and primitive ice-making devices were pioneered by a number of scientists including the aristocratic amateur, the Marquis of Worcester.

But understanding came only from the scientific attempts to comprehend the great developments in the practical uses of heat which were being made at the start of the nineteenth century. Heat had long been considered to be an invisible liquid-like substance called 'caloric'. This liquid could be pumped into materials or transferred from one object to another when contact was made between them. The motion of caloric was shown by changes in the temperature of the substances or objects being observed. The science of heat and energy will be more fully explained in Chapter 2; for the moment it is sufficient to say that the scientists of the first half of the nineteenth century discarded the notion of 'caloric' when they discovered that heat was really a form of energy, and that energy could also be found in mechanical motion, in chemical reactions, in electricity and magnetism. The transformation of energy from any one of these forms to any of the others was governed by quantitative laws. The chief problem in explaining, or understanding, these laws is that what actually happens in nature does not correspond exactly with our perception of what happens. The temperature of a body does not tell you how much heat it contains. Nor is the amount of 'heat' in a body the same as the amount of energy in a body.

The demolition of the caloric theory of heat was begun by the American Count Rumford, who pointed out in 1798 that when metal cylinders were bored out to make cannon, enormous quantities of heat were produced even though the amount of metal removed from the cylinder was very small. Indeed the amount of heat produced from the cylinder was apparently inexhaustible, and was quite incompatible with the current idea that 'inert caloric' was being squeezed out of the cylinder by the boring process. The exact meaning of Rumford's observation was further elucidated in the 1840s by the experiments of the Manchester brewer and amateur scientist, James Prescott Joule, who made careful measurements and showed that a certain amount of mechanical work produced a precise amount of heat. In other words, he showed that heat was a form of energy, and that heat energy could be compared with the energy of a moving body, or with chemical energy or electrical energy.

From Joule's time onwards scientists have thought about heat in this way: when coal is burnt in a furnace which is used to raise steam, which is used to drive a turbine, which is used to rotate a generator and make electricity, which powers an electric drill, which bores a gun barrel, from which the metal shavings are hot enough to boil a kettle to make a latter-day Count Rumford's cup of tea, then all along this chain of operations energy is being passed from one system to the next, either as heat or in the form called work.

The science of thermodynamics, developed by Joule and many others in the nineteenth century, was the study and measurement of what went on in a chain of processes such as we have described. The principal interest was in extracting useful work from a given source of heat. They were able to show what we now call the Second Law of Thermodynamics—that no engine designed for this purpose can ever be perfectly efficient, or in other words that some of the heat fed into any engine is always lost as heat to the surroundings, and one cannot get all the heat energy you supply out of the machine as useful work.

This principle is vitally important to refrigeration. The time came in the nineteenth century when engineers (and particularly people who wanted to preserve food) had a need to produce more cold than was provided by mixing salts and sal ammoniac with water or by volatilizing liquids such as ether. They had the idea of running a heat engine backwards, feeding work into it rather than taking work out, arguing that then the engine ought to suck in heat from the surroundings in exactly the opposite way to that in which it lost heat to the surroundings when it was run forwards. If this idea held good, then the surroundings of the engine would be cooled down. This is the principle on which all modern refrigerators work.

Our present-day low-temperature technology began to take shape in the 1850s when these concepts relating temperature, heat and energy could be applied to empirically observed phenomena. In the century that ensued our present urban, technological culture has been formed. It depends just as much on the power to make things cold as on the more obvious ability to make things hot. The roaring flame from the base of a Saturn rocket at Cape Canaveral convinces us that we can only reach the moon by burning fuels at extremely high temperatures and by providing materials in rocket motors that will stand up to these temperatures. It is not so obvious that the power to make things cold is equally important. But without liquefied oxygen, which is produced in extreme cold, the Saturn could never attain the power-weight ratio necessary in order to leave the earth's gravity. Nor could the crew survive in space. Without the comparatively modest temperature reduction in an astronaut's life support system, which keeps him cool enough to live despite the absence of a surrounding atmosphere to carry away his heat and sweat, men would be unable to walk around on the moon's surface.

But the exploration of the moon was no more than a demonstration of technological might, a political demonstration of industrial and scientific power in a battle of prestige between two nations. The real effects of low-temperature science and technology remain almost unnoticed at the roots of our urban way of life. A modern city, a Tokyo, a London, a New York, could scarcely exist without the ability to attain artificially temperatures of a sort that only occur naturally in the vilest weather near the Poles. For the vast majority of the people of these cities depend for their food supplies upon our ability to preserve food by freezing. The distribution of these food supplies depends in turn upon our ability to reach and maintain even lower temperatures. Our cities exist only because snow will delay the process of putrefaction of flesh, the phenomenon that so excited Francis Bacon that he caught his death of cold investigating it. The pattern of life that we know in those cities depends on low-temperature technology, too. The supermarket could not exist without the cold-cabinets that allow its meat, fish and vegetables to be pre-packed in cuts and quantities attractive to the housewife, and yet remain fresh.

The first important achievement of the low-temperature industry was, in fact, the shipping of frozen meat from Australia, New Zealand and the Argentine to Britain, and later to the industrializing countries of western Europe. Without this development neither the exporting nor the importing countries could have developed in the way they have.

And on the farms that supply the food for freezing, the ability to achieve low temperatures has wrought perhaps even greater changes. Anyone who drove through the English countryside in 1945 would have noticed that the cows were mostly brown and white in colour. A similar drive today would show a predominance of black and white cows. No breeding programme using natural means could have achieved this. It has been done by artificial insemination, replacing the old Shorthorn breed with the black-and-white Friesian cows which produce so much more milk. Artificial insemination on such a scale can only be achieved by freezing and storing sperm.

In our own homes our way of life has been changed and our standards of living have been improved out of all recognition in a single life span by low-temperature technology. Anyone of middle age has seen the movement from ice-box or stone slab in the larder to domestic refrigerator to deep-freeze. This is also the most homely example of the trend to lower temperatures—the domestic application of 'the colder the better'.

Many of the clearest ways out of the fuel and energy crisis are only open to us because of the ability to reach low temperatures. Just as low temperatures allow us to preserve food and thus to rationalize and organize its distribution, so they allow us to liquefy gases and achieve easier storage and cheaper distribution. It is impossible to send natural gas from Africa and the Middle East to the USA by pipeline. Yet the oilfields of the Persian Gulf and Nigeria produce huge quantities of natural gas, which were wastefully burned while the Middle West and central states of the USA suffered from gas shortages. Transport of liquefied natural gas in huge, refrigerated tankers is one answer. Similarly liquefied natural gas will be going to Japan and Borneo and the Middle East by tanker. The development of offshore oil fields, from Australia to the North Sea and the coasts of the North American continent, depends on the ability of divers to perform a comparatively small number of absolutely vital operations underwater. The divers in turn depend on gas supplies that can be made available on exploration rigs and production platforms far away from land. These bottled gases are supplied in the last analysis by the ability of modern refrigeration plants on land to achieve low temperatures at which the gases can be separated from each other, liquefied and transported to the scene of action.

The separation and liquefaction of gases is indeed the most important industrial application of low temperature technology. The scale of the operation and its results can be gauged from the simple statistics of a typical plant. In any one hour it will take in 8,000 cubic metres of ordinary air—that is the amount of air found in a theatre which can seat 1,000 people. The entire oxygen content of this air emerges as just two cubic metres of liquid oxygen, and this appears at one outlet of the plant. Other outlets produce neon, argon, krypton and xenon, all of which have uses of their own in industry—neon for lighting, argon for welding, and so on. The largest outlet of all at such a plant provides a flow of pure nitrogen. Barely ten years ago much of this nitrogen would have been allowed to return to the atmosphere. But now liquid nitrogen has become an eminently marketable commodity as it is the coolant of choice in the latest food-freezing techniques.

Liquid nitrogen is the coldest substance in industrial-scale production. The temperature at which nitrogen, the major component gas of the atmosphere, liquefies is minus 196°C. Following the rule 'the colder the better', the use of liquid nitrogen, both as a coolant and as a storage refrigerant, is rapidly increasing. Nitrogen has many advantages over other gases, notably that it is not explosive and is much less likely to react with other substances. It is quite common, nowadays, to find things being kept frozen in tanks cooled by liquid nitrogen—blood banks at transfusion centres and sperm banks at animal breeding centres are the most striking examples.

But apart from the storage of blood and some other tissues such as the cornea of eyes, the use of low-temperature techniques and technologies has not spread very far into the field of medicine. For this reason the field of cryomedicine is widely expected to be one of the most rapidly expanding fields in the immediate future. (The prefix cryo- is derived from Greek and is used to signify 'very low temperature'. Such words as 'cryobiology' and 'cryosurgery' are now in regular use and the word 'cryogenics' describes the field of science and technology which uses temperatures below minus 30°C; to cover the temperature range from minus 30°C to 0°C we use the terms 'refrigeration' and 'low temperature'.)

Cryosurgery is often called bloodless surgery because the extreme cold seals off blood vessels. But it is also quite common to use extremely cold substances—drops of liquid gas, for instance—to destroy a small number of cells, and this extremely delicate form of surgery has begun to be used in the treatment of many illnesses, notably of the brain.

This very ability of extreme cold to destroy living cells, which can be put to use surgically, is also one of the chief stumbling blocks in the application of low temperatures to medical progress.

What is needed is a method of storing whole organs or blocks of living tissue in such a way that they can be kept for long periods and then resuscitated, undamaged, when needed to supply a transplant. Large teams in some of the most advanced medical research centres in the world are working at this problem but they have not so far achieved finally satisfactory solutions.

Technologies that work below the temperature of liquid nitrogen, however, are still in the future. There are many who expect that liquid hydrogen will be the fuel of the long-term future. They look forward to the 'hydrogen economy', when huge quantities of hydrogen will be provided by using the electric power from nuclear power stations to break up water into its constituent elements, hydrogen and oxygen. Liquid hydrogen has already been used as a fuel in the American Centaur rockets and has successfully launched many satellites. The great advantage of hydrogen as a fuel is that the result of its burning is mostly water. The disadvantage which leads many to doubt the possibility of the hydrogen economy is the danger of explosions and fires—liquid hydrogen is much more inflammable than petrol. The extremely low temperatures involved—hydrogen becomes liquid at minus 253°C—may impose technical and economic barriers which are not worth surmounting on a large scale.

Helium is the last gas to liquefy as temperature is reduced—the liquefaction occurs at minus 269°C. Liquid helium is not in itself useful, for helium does not combine with other chemicals except under very unusual circumstances. But it is not too difficult to manufacture liquid helium which can then be used to cool other materials to this extremely low temperature. And, when brought to such very low temperatures, other materials, notably metals, display properties so extraordinary that entirely new fields of technology are opened up. The most valuable of these newly discovered properties is superconductivity, the state in which most metals and some other substances display no resistance whatsoever to electrical currents flowing in them. This immediately opens up the possibility of very much more powerful magnets and new types of electrical machines such as motors, pumps and generators.

The whole of the power-engineering industry will be revolutionized if we can bring superconductivity into general use. New types of power station can be built, and the transmission of electricity along superconducting underground cables could eliminate overhead lines and pylons. The crucial question is not so much 'Can it be made to work?'—there is little doubt about that, but whether we can make the refrigeration—the maintenance of heavy rotating machinery and miles of cable at temperatures below minus 250°C—economically viable.

Superconductivity has already been put to work in high-science technologies, notably in the construction of magnets of great power and comparatively small size for the world's largest particle accelerators (often called atom-smashers) where nuclear physicists are probing the ultimate structure of matter. But it is also used on a much smaller scale to provide magnets for comparatively small pieces of laboratory equipment such as machines for measuring nuclear magnetic resonances. The step from the laboratory and the special needs of scientists into the more rough-and-tumble world of manufacturing industry has not yet been taken, however, though the prospect is one of the most exciting features of low-temperature technology.

There are other strange phenomena that occur when temperature is taken below minus 270 degrees. Liquid helium itself exhibits 'superfluidity', the complete absence of viscosity: the liquid flows up the sides of a vessel, creeps over the rim and pours itself out. Strange reactions to the flow of heat and the behaviour of magnetic fields are also observed.

Physicists are very excited by these unexpected properties and unusual behaviours of matter at extremely low temperatures. The study of these phenomena is one of the frontiers of science at the moment. It is believed that some of the fundamental laws of nature—the laws of quantum mechanics—can best be studied by observing behaviour at the very lowest temperatures, and there is a continuing race to get to temperatures lower than any other laboratory has reached.

Eventually one reaches the concept of 'absolute zero'—the lowest temperature that can be imagined. This is minus 273°C, or 0°K. It is by no means uncommon for physics laboratories to record temperatures only one-millionth of a degree above absolute zero. Yet this does not mean that they are only one very small step away from the absolute, for the absolute zero is, by most definitions, necessarily unattainable. The concept of temperature, measured in degrees, becomes virtually meaningless. It is difficult to think of reaching some highest temperature, and likewise it is difficult to conceive any reality attached to the attainment of the lowest temperature of all.

In order to explain the strange phenomena that occur as materials are brought down towards absolute zero, the concept of internal order is more useful than the normal concepts of heat and temperature. Superconductivity, for instance, is easier to imagine if we think of the electric current as pairs of perfectly matched electrons slipping easily along between the rows and columns of perfectly ordered, motionless atoms of the metal conductor. At higher temperatures, with more energy in the system and less order, we can regard resistance as the trouble that a stream of excited electrons meets in fighting its way between ill-ordered lines of bouncing and vibrating metal atoms.

Although it is not likely that superconductivity will affect us in our daily lives in the very near future, there is one small area in which many of us are already reaping benefit from the ability to cool helium so much that it liquefies and can then cool other substances to the same extremely low temperature. This is in the use of satellites stationed above the earth for transmitting intercontinental communications. However powerful the electronics inside the latest satellites, the signals when they come back down to earth are extremely weak. To pick them up and amplify them back to usable strength, without distortion, special amplifiers have to be provided at the focus of the big dish-shaped antennae which are the most obvious features of the receiving stations on the ground. These amplifiers, called masers or mavars or parametric amplifiers, usually work at liquid-helium temperatures, and have to be kept in containers cooled by liquid helium. The principle is again that of taking advantage of the internal order of materials at very low temperatures. Because the material used in detecting and amplifying the satellite's signals is kept so cold, there is less generation of minute electrical signals caused by atoms interacting with each other. This 'electrical noise' of random interactions of atoms and molecules is present in all material, and increases with the internal energy. By cooling the materials we prevent the noise swamping the satellite's signals. So most of us who have made transatlantic or other intercontinental telephone calls have used liquid helium. We have also been helped by the trend towards ever-lower temperatures—'the colder the better'.

The whole of this vast movement into the realm of low temperatures has involved scientists in France, Poland, Germany, Holland, Russia, Britain and the USA to mention only those societies in which the most outstanding discoveries have been made. The pioneers of the technologies resulting from the scientific discoveries have also included Japanese, Australians and New Zealanders. The social effects of the technologies have been felt on a large scale in every developed country and in most other nations as well. But the lines of development that have led to superconductivity, satellite communications, space flight, the storage of animal embryos, and the possibility of banks of human organs awaiting the need for transplants, began in our appetites.

It was the desire for cool drinks in hot climates, the appetite for fresh and tasty foods, the need to preserve perishable meat, that started it all. The fact that so many people enjoy eating ice-cream was a major spur to the technology and manufacture of refrigeration machinery. The development of the fishing vessel that can freeze its catch and bring fresh-tasting fish to our tables is one of the factors that brought nations to loggerheads in trying to set up an international Law of the Sea.

The anthropologist and the philosopher interpret the myth of Prometheus as signifying man's creativity, either as technologist or as artist. But if an analogy can be drawn from the modern production of cold, then man first used fire, not to protect himself, nor to warm himself, nor to sharpen his weapons, but because he discovered the delights of roast meat. It is less edifying to think of one of our ancestors enjoying the taste of the carcass of some wretched animal caught in a bush fire, and deciding to burn his next captured prey, than to imagine the courage and conceptualizing power of a primitive man who decided to capture fire and bring it back to his cave for warmth and the protection of his family. But the less edifying may nonetheless be nearer the truth.

2 Temperature - And How to Cool It

We human beings are perpetually aware of being hot or cold. We spend a great deal of our time adjusting our surroundings so that we remain comfortable within the very narrow range of hot and cold that we enjoy—we put on extra clothing or take off our ties; we switch on the air-conditioning or turn up the central heating. Yet the human body does not have a special organ for perceiving heat in the way that we have evolved eyes to perceive light or ears to perceive sound.

Despite our preoccupation with heat comfort, this lack of a special heat organ means that our perception of heat is much less discriminating than our perception of other phenomena such as light or sound. Very often we cannot discern the direction from which the heat is coming; we can barely discern a difference in temperature between two objects external to our bodies, although we can measure the difference in temperature as several degrees. Our ability to judge whether one thing is hotter than another is notoriously erratic and can easily be deceived—a tin of orange juice and a bread roll that have been sitting next to each other in a freezer long enough to reach the same temperature are equally cold, but they certainly do not feel so.

We all observe in our normal experience that heat seems to flow from one object to another. An object can be heated by placing it in a flame. A piece of hot iron dropped into a flask of cold water appears to 'give up its heat' to the water so that the water gets hotter and the iron gets colder until eventually both are at the same temperature. Hot water poured into cold water appears to mix until we have warm water.

And so we have an intuitive concept of heat as a sort of liquid that can be concentrated, that can exist in objects and materials, and that can flow from one object to another, especially if they are in contact. Temperature should then be the measure of the concentration of this liquid—if we apply a large amount of heat to a very small object, that object becomes very hot or has a high temperature. We all continue to think of heat in this way. Our normal usage of the words 'heat' and 'temperature' continue to carry this concept implicitly.

This concept of heat and temperature is wrong. It does not correspond with the actual physical phenomena. And unless we realise that our normal view of heat is wrong it is impossible to understand the science or technology of low temperatures.

Heat must not be thought of as a liquid. Temperature must not be thought of as concentration of heat, nor is change of temperature a measure of change of heat. Under certain conditions heat can be pumped into a material without causing any change of temperature. Likewise heat can be removed from a material without causing a change of temperature. It is never strictly correct to speak of an object's 'containing a certain quantity of heat'. Temperature is no more than the measure of a certain characteristic of a material under certain conditions. Temperature can be changed without addition or substraction of heat—for instance, the temperature of a quantity of gas can be changed by altering only the pressure of the gas.

The early scientists of the seventeenth and eighteenth centuries naturally took their start from the intuitive view of heat, and developed the 'caloric' theory of heat described on p. 19. As we have seen this theory was disproved, but it has left behind it a serious problem of understanding. Because caloric theory was based on an intuitive view of nature, it used the words and concepts of this view—thus heat 'flowed' from one object into another. Although the theory has now been discarded, the words have been kept, and scientists still talk about 'heat-flow', although they do not accept that anything actually flows from one body into another. Unfortunately most of us still think in terms of flow of heat, or that high temperature corresponds with a great 'concentration' of heat, and this makes it difficult to get a clear understanding of the meaning attached by scientists to certain words in the framework of thermodynamics.

The first problem for the scientists, or early engineers, studying heat and trying to use it in machines, was to find quantities they could measure. The most obvious of these was temperature, for it seems at first as if change in temperature must be the same as change in heat. Many eminent men constructed scales of temperature, from Galileo to Sir Isaac Newton, who favoured the normal body heat of a healthy man as one of the fixed points. But it was an instrument maker, rather than the theorists, who first established a widely used temperature scale. Gabriel Daniel Fahrenheit was born in Danzig but he spent most of his professional life in Holland and England. He made mathematical and astronomical instruments of various sorts and in 1714 he turned his attention to the manufacture of thermometers containing alcohol or mercury as the measuring fluid in a narrow glass tube. As the fluid expanded or contracted in the tube it measured the increase or decrease of temperature.

The coldest environment Fahrenheit could create in his laboratory was in a bath containing ice, water and sal ammoniac. This he designated his zero point. He accepted the suggestion of Sir Isaac Newton that the temperature of a healthy human body—a standard widely available for calibration—should be another fixed point. He chose to call this 96 degrees, possibly because this number is divisible by 2 and 3 and 4. Since the freezing point of water came at a point a third of the way along his scale, that became 32 degrees. (The refixing of normal healthy body temperature at 98·6 degrees has come with greater refinement and accuracy over the years.) By extending his scale upwards, Fahrenheit established the boiling point of water as 212 degrees and equal spacings of temperature between his fixed points gave readings of degrees which defined the scale.

Fahrenheit's scale established itself widely. Other temperature scales were afterwards proposed, notably by Reaumur and Celsius, and Celsius's suggestion of fixing the zero point at the freezing point of water and calling the boiling point of water 100 degrees is nowadays prevailing over Fahrenheit's scale in English-speaking countries. But Fahrenheit's invention of significance was not his scale of temperature but his thermometers, which enabled scientists, wherever they were working, to read off the same temperatures from identical phenomena. Thus they could communicate with each other their measurements, the basic numerical material of all science, and measured actions or quantities of material could be equated with measured changes in temperature.

Roughly parallel with the development of thermometry was the development of a whole line of experiments which were really the precursors of modern chemistry rather than of physics and engineering. These were experiments on gases, started by the Anglo-Irish aristocrat Robert Boyle in the seventeenth century, and carried on by scientists in many other European countries, notably in France. These experiments showed that if a certain quantity of gas is put under pressure so that its volume is decreased, the gas becomes perceptibly hotter. But if the gas is heated first, it increases its pressure on its container and increases its volume. Boyle's Law, relating the pressure, volume and temperature of a gas, is one of the first learned by students of elementary physics.

Developments in this field led to the concept of the gas thermometer, theoretically the nearest to an ideal thermometer, though nothing like as simple, useful, reliable and practical as a Fahrenheit thermometer for measurements in the normal middle range of temperatures in which we live and in which eighteenth-century scientists had, perforce, to operate. The principle of the gas thermometer is that while the volume of a sample of gas is kept constant, the change in temperature of the gas is measured directly and exactly by the change in pressure that the gas exerts on its surroundings.

If we think about heating the gas up, we can imagine an ever-increasing pressure as the gas gets hotter and hotter; but if we think about reducing the temperature the gas will exert less and less pressure, until we arrive at a temperature so low that the gas will exert no pressure at all. It is ridiculous to think of a gas exerting a negative pressure on the walls of its container, so there must be a point beyond which the temperature can go no lower. There must be an end to reducing temperature; there must be an absolute zero.

There was a hint of this concept in the work of Amontons in the seventeenth century, but it was not until the following century that two French scientists, Charles and Gay-Lussac, demonstrated the same conclusion quite independently. They started with a volume of gas at 0°C, the temperature of freezing water (or melting ice), and measured its pressure. They could not lower the temperature of the gas very much with their simple mixtures of ice and salts, but their results were perfectly clear.

Both men found that the pressure of the gas was reduced by $\frac{1}{273}$ of its original value for every degree Centigrade that the temperature was reduced. The implications were twofold; first, and most important at that time, was the establishment of the accuracy both of the gas thermometer and of the laws relating to temperature, pressure and volume; second, they established that the pressure of the gas would be nought at minus 273 degrees. Absolute zero, therefore, must be minus 273 degrees, the temperature below which it was impossible to go. This is in fact the temperature we still accept as absolute zero. In much modern science this is called 0° Kelvin (K.), and on the Kelvin scale the temperature of melting ice is 273°K.

But it was this type of experiment that showed scientists that temperature is not directly related to heat. If we bring a kettle to the boil we can measure the temperature of the boiling water as 100°C. or 212°F. If we keep the electricity switched on, or the kettle over the fire, the temperature will not rise appreciably but the water will all turn to steam eventually and the kettle will boil dry—putting in further heat has not changed the temperature of the water, it has changed its state; it has turned it from liquid water into steam or water vapour. And we can increase the temperature of a gas simply by applying pressure to it—by compressing it without applying any heat.

What then is temperature? What do thermometers measure? Some physicists answer this question, perfectly reasonably, by turning it on its head and saying that the thermometer reading is what defines temperature. The alternative answer, that temperature is the degree of hotness, gets little further. The situation is most easily understood by reference to the traditional gas-in-a-container experiments. The gas is a collection of atoms, or molecules which are simple combinations of atoms. The atoms are all in motion, whizzing about, banging into each other and crashing against the sides of the container. The combination of particles hitting the container are seen by the observer as the pressure exerted by the gas on its container. If the particles are made to move faster in the confined space they crash against the container walls faster and more often—which we see as increased pressure. We therefore say that the increased temperature means that the particles are moving faster. And more accurately, we find that the increased temperature is related to the increased kinetic energy of the particles, the increased energy of their faster movement. Temperature can therefore be thought of as a measurement of the speed of movement of the atoms and molecules that make up all matter. A gas, however, is the state of matter in which the atoms are not ordered in positions; they are free to rush about. A liquid is a more ordered state of matter and a solid is a state of matter in which atoms remain in their ordered positions. It has turned out, as scientific knowledge has deepened, that atoms can vibrate about their 'true' or 'proper' positions in the ordered structure of a solid, and this vibratory movement requires energy and therefore can be measured as a temperature. Furthermore, atoms can vibrate about their position in a molecule, and whole molecules can vibrate in themselves and about their ordered positions; all these vibrations appear as temperature when we measure the degree of hotness of a sample.

Physicists and engineers have, of course, equations, proved by experiment, expressing the precise relationship between temperature read by thermometer and the energy of movement of the atoms in the sample. The normal sets of relationships expressed by these equations cease to apply when the temperature of absolute zero is approached, for here the energies involved are so small that they can only appear as discrete small packets—quanta—and the investigators are in the realm of quantum mechanics. It is precisely because this is a region in which they can perform experiments which shed direct light on to the theories of quantum mechanics that scientists are so interested in the area near absolute zero.

This explanation or 'visualization of the meaning' of temperature also shows why temperature is not a measure of heat. Heat is one form of energy along with chemical energy, electrical energy and mechanical energy, all of which can be transformed into each other. The heat energy put into, or taken out of, a sample of material can not only affect the movement of the atoms which is measured by temperature, but can also affect the chemical or electrical forces of attraction or repulsion between particles and molecules—and this in turn is seen as the orderliness or otherwise of the arrangement of atoms in the sample. One can never correctly speak of the amount of heat in a body, although it is perfectly possible to measure or calculate the amount of heat that has been put into or taken out of a body in some operation. In particular, although absolute zero may be conceived as a state of complete orderliness and absence of motion, it does not follow, even theoretically, that it is a state with no energy at all.

The development of this understanding of the nature of temperature and its relation to heat was not purely academic. It was a necessary step in discovering ways to produce cold. Whereas it is possible to think up techniques for producing greater heat—moving from wood to coal for fuel, introducing greater draught and airflow, and so on—it is not possible to guess how to produce greater cold.

The obvious way to cool something is to place it in contact with something colder. Then the two will share their kinetic energy, although this implies that the cooler object will warm up at least some of the distance towards the original temperature of the hotter object. But by collecting enough ice or snow and allowing some to melt it is possible to bring temperatures down to 0°C easily enough.

Fahrenheit and other early scientists succeeded in lowering the temperature of water below its normal freezing point by using mixtures of ice, water and salt, or ice, water and sal ammoniac. The reason these worked was quite subtle, not intuitively obvious, and was discovered by the chemist, Black, who studied the melting of large blocks of ice in the last years of the eighteenth century. He published his work in 1803 and the relevant portion reads, 'The opinion I formed from attentive observation of the facts and phenomena is as follows. When ice, for example, or any other solid substance, is changing into a fluid by heat, I am of the opinion that it receives a much greater quantity of heat than what is perceptible in it immediately after by a thermometer. A great quantity of heat enters into it on this occasion without making it apparently warmer when tried by this instrument. This heat, however, must be thrown into it in order to give it the form of a fluid; and I affirm that this great addition of heat is the principal and most immediate cause of the fluidity induced.' Black had thus discovered what we now call latent heat, the fact that considerable quantities of heat must be applied (or taken away) for a substance to change between the states of solid, liquid or gas, and that the temperature does not change during this change of state although heat is being moved.

At the time of Black's work, his results were interpreted as showing that liquids mopped up caloric fluid and rendered it inert to an even greater extent than solids. But Humphrey Davy showed that two pieces of ice could be melted simply by rubbing them together, which rendered this position untenable. We have seen, in Chapter 1, that Joule finally provided an alternative explanation in the 1840s—that heat is another form of energy. We now understand that the heat in a substance is mostly stored as the energy of motion of the molecules in the substance, either as they travel about or as they vibrate to and fro about their 'true' positions and the temperature of the substance is the expression of the speed of motion of the molecules.

But there is also another form of energy locked up in a substance. This arises from the fact that if any two molecules are close enough to feel the mutual force of attraction that makes matter cohere, then energy must be supplied to force them apart. This energy, stored within the substance, is called potential energy. Because the molecules of a liquid are rather further apart from one another than they are in the corresponding solid, the potential energy of the liquid is different. However, if a liquid and a solid are in contact at the same temperature, as in the case of blocks of ice floating in water at 0°C, the molecules will be travelling at exactly the same speeds, and will therefore have the same energy of motion, and are therefore showing the same temperature. But there is a difference in the potential energy of the molecules in water and in the ice, and it is this difference that accounts for the phenomenon of latent heat.

The latent heat of melting is relatively large, and that is why ice is an efficient coolant. If 100 grams of ice-cold water at zero degrees (0°) are stirred into 80 grams of boiling water at 100 degrees, the temperature of the final mixture will be about 44 degrees—rather hotter than your hand can bear. But if 100 grams of ice at 0 degrees are mixed with 80 grams of boiling water, the result will be 180 grams of water at a temperature only just above the zero degrees (0°) of the original ice. The difference lies in the large latent heat of melting of the ice, that large potential energy which has to be given to the water molecules to separate them from their close mutual attractions in the solid state.

Though the molecules of a solid separate to some extent when the solid melts, they separate much more completely when the resultant liquid evaporates and turns into a gas. We might reasonably expect therefore to find another latent heat, that of vapourization, and this should be even greater than the latent heat of melting. This turns out, indeed, to be the case; the latent heat of vapourization of water is seven times greater than the latent heat of melting ice. If you switch on an electric kettle full of ice at nought degrees it will take about three minutes for the ice to melt, about three and a half minutes to heat up to boiling point and about twenty minutes to boil dry. And just as the temperature stays at nought degrees during the first three minutes while all the heat supplied by the filament is being taken up as the latent heat of melting, so the temperature remains at 100 degrees throughout the final twenty minutes when all the heat is being taken up as the latent heat of vapourization or evaporation.

As men have long used the latent heat of melting ice to cool other substances, can we use the much greater latent heat of evaporation to achieve even more efficient cooling?

We need, for a start, a liquid which is more volatile than water. Ether was used in the early days of refrigeration, and it achieved its effect of cooling in precisely this way—by taking heat from the surrounding warm objects or environment as the latent heat needed for its evaporation. But the liquid we use most commonly in modern domestic refrigerators is freon; its chemical formula is CCl_2F_2. Under normal conditions of temperature and pressure, freon is so volatile that it is a gas, but it can easily be turned into a liquid by compressing it, and then it can be stored in cylinders as a liquid, just like the butane which we regularly use as 'Calor gas'.

Imagine that you have a cylinder from which all air has been removed and which is half-full of liquid freon; then what fills the other half of the cylinder? The answer is freon gas, the vapour formed from the liquid. This gas is at a pressure that depends on the temperature of the cylinder—it will probably be about six times as great as atmospheric pressure if the cylinder is in a normally heated room. If the valve of the cylinder is opened the freon vapour will flow out into the room and its pressure inside the cylinder will drop sixfold. The freon liquid will then start to evaporate in its effort to restore the vapour pressure to its original value. It will need latent heat to achieve this evaporation, and if extra heat is not immediately available to supply this the liquid will cool down and so will the cylinder around it as the heat is drawn in.

If you performed this operation, leaving the cylinder valve

open to the atmosphere, the cylinder would cool down to about minus 30°C. This is the standard boiling point of freon—which means that it is the temperature at which the vapour given off by the liquid freon exerts a pressure which is sufficient, but only just sufficient, to balance the pressure of the atmosphere. But if you attach to the cylinder a pump which will suck away the freon gas as it evaporates and which will maintain a vapour pressure less than one atmosphere, then the temperature of the liquid (and the cylinder) may fall still further as more heat is sucked in for the latent heat of evaporation of more freon. The limit to the further lowering of temperature is set only by the speed with which the pump can suck away freon vapour and the extent to which the cylinder can be insulated to prevent heat being sucked in from the surrounding environment. In practice you could probably achieve a temperature as low as minus 100 degrees with such a system.

A domestic refrigerator cannot, of course, afford to pump away all its freon. So there is a system provided which will compress and reliquefy the freon vapour which has been pumped away. The liquid is fed back through a valve into the evaporation chamber. So the freon goes round and round a closed cycle, evaporating at low pressure in one place and condensing at higher pressure in another part. Just as it cools and draws in latent heat from its surroundings while it is evaporating, so it warms and throws out heat when it is condensing. The machine is arranged so that it draws in heat from the inside of the cabinet and the food it contains, and throws out heat to the rest of your kitchen. The freon is normally driven round the cycle by an electric motor, and the energy which is discharged into your kitchen includes what has been drawn from the electricity supply as well as the heat taken from the food.

Another common type of domestic refrigerator—the vapour absorption, or 'Electrolux' type—uses ammonia as its circulating fluid. It does not need a pump and compressor; it simply uses heat to drive the ammonia round, mixing it as a liquid with water and as a vapour with hydrogen. The advantages of the vapour absorption type of refrigerator are that it can use different energy sources and has no moving parts; but it cannot compare in efficiency and cooling power with the freon or 'vapour compression' system of refrigerator.

The commercial development of refrigerators began in the second half of the nineteenth century for reasons that will appear in the next chapter. At that time substances such as freon were probably unknown and certainly unavailable. It was many years, therefore, before the vapour compression refrigerator came to dominate the commercial market. And before it did so, two other methods of cooling were proposed by the famous Scottish physicist, Lord Kelvin. Both played important parts in the scientific search for ever-lower temperatures, and one of them took on industrial importance; this we shall look at first.

Anyone who has used a bulb of compressed carbon dioxide—a sparklet—to make soda water will remember how the bulb cools and often forms 'frost' on its outside when it is punctured. This occurs because, as the gas suddenly expands and pushes back the outside atmosphere, it uses the reserves of kinetic, or motion, energy which have been put into it by the compression process and which have been trapped inside the bulb. Exactly the same thing would occur if we used the pressure in the carbon dioxide to drive a piston instead of drive the atmosphere. But driving a piston is 'doing work' and could be incorporated in some machine. If we drove the pistons of a machine with compressed carbon dioxide released from bulbs we should be supplying the machine with energy drawn from the reserves of kinetic energy in the compressed gas—and we should be cooling the gas.

Scientists call this type of cooling—the cooling of an expanding gas—adiabatic expansion. Kelvin suggested that it might be used in a refrigerating machine that contained air as its working, or circulating, substance. His idea was that the air would go round and round a closed cycle, being compressed by a piston at one point before being allowed to expand, and cool, at another.

Kelvin's idea was turned into a practical machine by a technician called James Coleman, who worked for the Glasgow engineering firm of Bell. The machines were called Bell-Coleman refrigerators and, though they are no longer used today, they played a vital part in the history of our modern urban civilization, as we shall see in the next chapter. But quite as important was the principle of cooling involved. If the machine is arranged so that the heat of compression is carried away by a flow of some cool substance, such as water or cooled gas, the system can be made to work so that there is a steady cooling of the circulating substance, air or another gas, by a small amount on each circuit of the system. This concept, though not very efficient in practice, has played an immensely important role in the science of reaching very low temperatures.

Kelvin's other important idea about cooling derives from the time when he was still plain William Thomson and was carrying out some experiments with Joule. They showed that most gases cool slightly when they are allowed to stream slowly through a very small nozzle or through the many small pores of a porous plug, where the pressure downstream of the constriction is rather less than it is above. The effect is a very subtle one, and it is only by detailed analysis in the modern style that it is possible to tell whether the gas will gain more or less energy from the compressor pushing it along than it uses to push back whatever is in front of it—normally more of itself at lower pressure. Correspondingly the cooling effect is very small, and it differs from one gas to the next. But the Joule-Thomson device proved very important to scientists of the late nineteenth century when they had reached the limits of cooling by expansion.

While all these practical advances in refrigeration had been developing through the nineteenth century, there had also been developments in those engines which used heat to perform useful work and provide motive power for the myriad machines of the Industrial Revolution. Theoretical science had lagged behind the engineers and technicians and had struggled along trying to provide explanations of what the practical men had achieved.

They had set about their task, in the absence of any satisfactory theory of heat or energy, by measuring the things they could measure and seeing what relationship held between the quantities they could measure. Thus they built up the Science of Thermodynamics. This essentially developed two great laws. The First Law of Thermodynamics states that energy can never be destroyed or created, it can only be transformed into different forms of energy. Hence it is also called the Law of Conservation of Energy.

The Second Law of Thermodynamics was pioneered by a French Engineers officer, Sadi Carnot, who founded the whole theory of heat engines that we still use today. He published his *Reflections sur la puissance motrice du feu* in 1824, and though it did not contain the final formulation of the Second Law of

Thermodynamics (which is credited, rather later, to a German, Rudolf Clausius), it established the most practically useful deduction from the law, which is that we can get more useful work out of a machine if we increase the difference in temperature between the hot parts of the whole cycle and the cold parts.

When the Second Law was formulated it made its appearance as a mathematical equation—an equation which both mathematicians and physicists find very satisfying but which inevitably puzzles the layman. The equation shows what we have already noted, that you can never make complete use of all the heat energy you put into a machine, some is always wasted. In one quite useful sense this shows that you can never make a perpetual motion machine. But the 'term' of the equation which speaks of this wasted heat does not make it obvious what is going on, nor does it speak directly of wasted heat. It is, instead, a term describing something which cannot be seen by ordinary observation. The early users of the Second Law called this mysterious quantity 'entropy'; they did not understand it properly, but all measurements of actual heat engines at work showed that it existed. And the nature of the equation was such that whatever system was used, in the long run the entropy always increased as the result of any operation. In the ordinary sense that means that there is always some heat wasted, but it became clear that there was a philosophical meaning to entropy as well. If entropy is always increased overall by any operation, then eventually the entire universe will run down until it consists only of entropy.

Nowadays we understand entropy more clearly. It can be seen as a measure of disorder. If at any stage of an operation entropy is decreased then greater order has been imposed on that part of the system—although this will eventually have to be paid for, according to the Second Law of Thermodynamics, by an even greater increase of entropy over the entire system. Entropy, in the visualized sense of order, becomes very important in low-temperature operations. When a specimen of any substance is at a very low temperature it is usually a solid, a crystal, with the atoms or molecules all lined up in proper positions, moving very little, vibrating very little. In other words we can regard making something very cold as imposing a very high degree of order on it, or decreasing the entropy.

Scientists nowadays use the concepts of entropy, order and disorder to plan ways of reducing the temperature of substances ever nearer to the point of 'absolute zero'—which can also be regarded as the point where perfect order is attained.

But it has taken 2,500 years to reach the stage where we can even contemplate the concept of absolute zero—and incidentally reach the point where we know we can never achieve this final temperature. During most of this time science has been struggling along trying to explain the results produced by practical men. It is only in the last hundred years that scientists have taken the lead and showed engineers how to achieve lower temperatures. And during most of these long centuries, even during most of the last hundred years, the driving force behind the practice and theory of refrigeration has most often been the human desire for cool drinks, fresh food and ice-cream.

3 Food

Alexander the Great may have known about ice-cream, for he is recorded to have enjoyed a delicious mixture of milk and fruit juices, sweetened with frozen honey, and cooled by ice and snow brought from the mountains. Egyptian pharaohs, Roman emperors and Greek tyrants are all reported to have cooled their wine with specially imported snow. Something of this sort must have been very popular and enjoyable because the earliest doctors are to be found inveighing against the practice. Hippocrates, for instance, the physician from the island of Cos in the Aegean, the man who formulated the Hippocratic Oath, wrote in 460 BC: 'It is dangerous to heat, cool, or make a commotion all of a sudden in the body, let it be done which way it may, because everything that is excessive is an enemy of nature. Why should anyone run the hazard in the heat of summer of drinking *iced waters*, which are excessively cold, and suddenly throwing the body into a different state than it was before, producing thereby many ill effects? But for all this, people will not take a warning, and most men would rather run the hazard of their lives or health than be deprived of the pleasure of drinking out of ice.'

In all reasonably sophisticated societies in warm climates, from Portugal and Sicily to Egypt, Persia and India, there were methods of cooling water and even of freezing it artificially. Porous pots, which allowed water to percolate slowly outwards and then evaporate, were most common. Mixtures of salts, nitre and saltpetre with water were known by observation to have temperature-lowering effects. Even in the steamy heat of Bengal the Indians could make ice on the surface of shallow, reed-lined pits exposed to the night air.

It is more than possible, however, that the Chinese used iced drinks and iced confections long before the Mediterranean peoples. When Marco Polo, the Venetian traveller, returned to Europe in 1292 from his visit to the Mogul Emperor of China, Kublai Khan, he brought with him a recipe for ice-cream. He had found this mixture of frozen milk and fruit-juices being sold from handcarts in the streets of Peking, and the Chinese assured him they had known about it for 3,000 years. From Italy to Paris the recipe for water-ices or 'sorbets' came in the train of Catherine de Medici in the first half of the sixteenth century. And still the doctors deprecated such delicacies. The physician Champier was in the Court of Francis I when that glittering French monarch went into conference with the Emperor, Charles V, and Pope Paul III at Nice. The French doctor was astonished when he saw the Italians and Spaniards put snow, which they had specially ordered to be brought from the mountains, into their wine to cool it. He immediately condemned the practice. This did not stop the French princess Henrietta Maria bringing the recipe for ice-cream to England when she came to marry Charles I in 1630. That unhappy king paid the queen's French chef a pension to keep the recipe a secret. It was not until 1769, however, that the phrase Ice Cream first appeared in literature, according to the Oxford English Dictionary, nor had *glacière* appeared in French dictionaries much before the end of the seventeenth century, though iced fruit drinks were sold by Parisian *limonadiers* throughout the second half of that century.

The storage of ice throughout the summer in ice-cellars or half-buried ice-houses seems to have been practised in Constantinople in the late Middle Ages at least. A French traveller, Ballon, suggested that his countrymen should take up the idea which he had observed in 1553. By the end of the sixteenth century some such devices were almost certainly being used in France, for the effeminate luxury of the Court of Henry III included the use of snow to cool drinks in midsummer. The idea spread to England in about 1650, and from 1750 onwards the construction of ice-houses in the grounds of great mansions became normal—so much so that these ice-houses are now considered part of the British architectural heritage and are preserved by the National Trust in several instances.

The history of ice-cream includes an interesting episode at what is believed to be the first serving of iced dessert at the White House in Washington, during the Presidency of James Madison. Mrs Dolly Madison had ordered custard-pies, frozen in the ice box, for a dinner. One of the lady guests was so terrified on taking her first bite that she screamed out, 'Poison!'. The cook was actually seized by security guards before Mrs Madison quelled the panic by eating one of the frozen sweets herself and declaring it delicious.

But ice, as a pleasure and as a preservative, was known to the Americans some time before this incident. The Royal Governors of Virginia built an ice-house at Williamsburg. George Washington supervised the collection and storage of ice from his ponds at Mount Vernon. Jefferson designed an ice-house for Monticello. The pattern of European society had traversed the Atlantic, and the ice-house, as a feature of luxury, was part of the great house on its estate. But somewhere around 1800 a new pattern started to emerge as ice became more widely used, and therefore a marketable commodity. Ice moved down the social scale and became useful to the middle-class housewife and the working farmer. It no longer remained a luxury used only by the most wealthy. Ice began to be used to keep food fresh as well as to provide the sensation of cold in hot climates.

Not surprisingly ice was in greatest demand in the Southern States of the USA, while the best sources of ice remained in the North, in New England. Charleston had an ice market in the streets by 1799. Sea captains sometimes ballasted with ice for their empty runs south from Boston via New York and Philadelphia to the Carolinas. With remarkable speed all these factors were put together and the Boston ice trade began. In a Boston graveyard there is the tomb of William Fletcher who died in 1853 at the age of 83. His monument records, 'He was the first man that ever carried ice into Boston market for merchandise.' If he did this as a go-getting young man it must have been well before 1800.

And in 1803 we have the first use of the word refrigerator. It was not referring to what we now call a refrigerator, which is a machine for producing cold; it was in fact an ice-box. But whereas to the Englishman the cold-producing machine in his kitchen is always 'the fridge', we notice that many Americans still call their refrigerator 'ice-box'.

This first 'refrigerator' appeared in *An Essay on the Most Eligible Construction of Ice-Houses; also a description of the newly invented machine called the Refrigerator*, written by Thomas Moore, who farmed at Brookeville in Maryland, twenty miles from Georgetown, which is now Washington, DC. After describing his ideal ice-house, a nine-foot-deep pit on the north side of a hill, lined with wood, roofed, and provided with a raised floor to allow for drainage of melted ice, Moore comes to his refrigerator. This was a box of cedar wood with a tin vessel nailed to the lid in such a way that the contents of the vessel could be packed round with ice. Outside there was further insulation by rabbit fur, pelt inwards, and cloth. With this refrigerator Moore could carry 22lb of butter to market where it arrived in such fresh condition that it commanded fourpence to fivepence halfpenny more per pound than anybody else's. Thus his refrigerator paid for itself in four journeys. Moore patented his idea and offered the public the opportunity to take out licences to make similar refrigerators. This produced virtually no result, and it was his writings that had the greater effect. According to the historian of the early ice trade, Richard O. Cummings (1949), 'His pamphlet was for many years the most important work on refrigeration in relation to transportation.'

Moore's pamphlet also demonstrates clearly that the use of ice had ceased to be simply a luxury of the rich and had become a commonly practised method of preserving food, or at least keeping its freshness, in the daily marketplaces.

The real rise of the Boston ice trade began very shortly after Moore had published his pamphlet. It was in 1806 that a Boston lawyer, William Tudor, started his attempts to organize a large-scale ice-exporting business. His first efforts were aimed towards Havana and the British and European owned islands of the West Indies. He sought to obtain monopolies for the supply of ice to these places, and thus he hoped to displace the random voyagings of Boston sea-captains carrying ice as ballast on their southward journeys to collect vegetables and fruit from the market gardens and small farms of the New York and Philadelphia areas and trade them south to the Carolinas and the Caribbean, or north to New England. Tudor also tried to start trade in New England fruit by attempting to carry it, preserved in ice, to New York.

The Anglo-American War of 1812 interrupted this promising start, but when hostilities were over the trade picked up again. Tudor heard in 1816 of a New York trading captain who took beef southwards packed in ice, and sold not only the beef, but also the ice in Charleston, where it fetched 12½ cents per pound weight. Tudor was still persevering with the West Indies trade, where his monopoly at Havana brought him 25 cents per pound for ice. Competition however flourished on the mainland. Ice carried to Georgia fetched only 6 or 8 cents per pound. Tudor's brother, Frederick, who took over the business, set up ice depots at Savannah and New Orleans, but in the latter port he faced rivalry from merchants who floated ice down the Mississippi. Ice had become such an important part of life in the Southern States and the Caribbean, however, that when the warm winter of 1818 caused a shortage of ice in New England, the captain of the brig *Reprieve* found it worth his while to sail up to Labrador, find an iceberg, attack it with axes and saws and bring the pieces back south. By the mid-1820s, Tudor alone was exporting 2,000 tons of ice a year from Boston and that city's total export of the commodity was about 3,000 tons.

Furthermore, the vigorous export of natural ice had already had the effect of slowing up, if not completely halting, the manufacture and sale of ice-making machines. The use of evaporating ether to turn water into ice was well known before 1800—Benjamin Franklin was one of those who wrote about it. Another 'philosopher', Gerald Naime, used sulphuric acid to absorb water vapour, thus giving the effect of pumping off the vapour. This pumping away of the vapour from above a liquid causes a reduction in the temperature of the liquid because the liquid gives up energy to the formation of more vapour (see Chapter 2). Household ice-making machines using sulphuric acid were actually manufactured in Britain and exported to the West Indies, but they failed to hold a market in face of cheap Boston ice. Similarly, the proposal to build an ice-making machine in New Orleans, using a steam engine to pump ether, came to nothing. This proposal was put up by Oliver Evans, a steam engineer of Philadelphia who published the strangely titled pamphlet, *The Abortion of the Young Steam Engineers Guide*, in Philadelphia in 1805.

The competitiveness of the Boston ice trade was vastly improved in 1825 when Nathaniel Jarvis Wyeth invented a new device for harvesting ice. Wyeth was a descendant of one of the very earliest New England settlers, and a native of Cambridge, Massachusetts. He was a hotelier and he owned orchards and land on the banks of Fresh Pond just outside Boston. His invention was really a development of the metal runners on a sledge. He gave one runner a series of cutting edges, successively deeper towards the rear of the runner, which cut grooves in the ice when pulled across the surface by a horse. The second runner was plain, but it was made to travel in the groove cut by the first runner in the previous passage across the ice. A series of passes in one direction, followed by a series of passes at right angles, thus left a pattern of squares or rectangles cut deeply into the surface. Once one of these squares had been broken up it was found that the ice readily split off along the lines of the other cuts. Thus blocks of ice of regular shape were produced, which meant, in turn, the ice could be stacked, retained by planking and properly insulated with hay. Thus the work of harvesting could be further mechanized, and huge ice-stores could be built on factory patterns, with moving belts and steam-driven chains equipped with huge hooks to move the ice around for organized marketing and distribution. Tudor took up Wyeth's invention and took him into partnership, and their application of the new device reduced ice-harvesting cost from 30 cents to 10 cents a ton. The regular shape of the ice also enabled them to start selling ice by weight rather than by volume measures of ice chippings, which had been the practice up to that time.

Wyeth went on to develop a new horse-drawn chain-saw which further improved the efficiency of ice-harvesting. Tudor, in his diaries (which are in the Baker Library at Boston), records a meeting with Wyeth in the winter of 1828 when the second machine was being developed. 'January 3rd, 1828. I found Wyeth wandering about the woods at Fresh Pond in all the lonely perturbations of invention and contrivance. His mind evidently occupied in improving the several contrivances which he is perfecting for carrying into good effect improvements in his several machines for the ice business. I have, from time to time, given him several hints, particularly respecting the ice-cutter, which I first suggested to him, and he has improved the plan of last year, and now tells me that he has improved upon this year's improvement.' One can see the seeds of later business quarrels in this passage, and these disputes split up the Tudor-Wyeth partnership after Wyeth had been granted his patents in 1829.

But from this time the ice business developed into a fiercely competitive affair, which later evolved into price-wars and rivalry between delivery services. Especially when a warm winter threatened normal harvesting of the usual quantities of ice, expeditions would go out with gangs of labourers to more distant ponds or up the Kennebec River in Maine. Work might go on all night by torchlight, and the courts had to decide the rights of ice between competing owners of lakeside land. Ice harvesting was also becoming a major occupation in other states, notably in upper New York State, where the Hudson River was an artery for ice transport.

Fresh Pond itself retained a primacy among ice-fields for many years, not only because of its closeness to the port of Boston but also because its ice was remarkably clear. This clearness in ice was a much-sought-after quality, probably because the consumer thought that it implied purity. When the first ice-making machines began operating on a large scale, their chief marketing problem was the bubbles in their product. Twenty years after Wyeth's inventions revolutionized ice harvesting at Fresh Pond, a US company began bringing ice from Norway into England. From its original icefield at Wenham Lake, near Salem, Mass., this company was named Wenham. The word Wenham for a time became a synonym for clarity and purity in England. (Fresh Pond is now, however, incorporated into Boston's water supply system.)

Frederick Tudor ran into business vicissitudes in the 1830s and left the ice trade and emigrated to Oregon. Addison Gage of Charlestown, a forceful livery-stable keeper, entered the ice trade from the delivery side and became the leading ice-merchant of Boston after Tudor. In the 1830s Bostonians were shipping ice to South America, Egypt and Gibraltar, as well as to the southern parts of their own continent. There were regular ice-depots run as separate businesses in most major southern cities of the USA, supplied by men in the north who specialized as ice-merchants.

In 1833 a huge new area of trade was opened up when Samuel Austin, a New England East India merchant, decided to open an ice-depot in Calcutta, then the commercial capital of the British Empire in India. He chartered the ship *Tuscany* and it left Boston in the autumn of 1833. After twice crossing the Equator it landed its cargo of 120 tons of ice (provided by Tudor's company) in Calcutta. In the years immediately following at least six other Boston merchants shipped ice to Bombay and Calcutta, and as an offshoot of this trade, Boston ice was even landed in Australia. This was followed by the shipment to India of perishables such as apples, butter and cheese, packed in ice. Fish packed in ice was now going from New England to the West Indies, along with other perishables, and by the late 1840s eighteen cargoes a year of perishables, each cargo worth 2,500 dollars, were being shipped from New England to India and the West Indies.

By this time the internal consumption of ice in the rapidly expanding US domestic market had made such farflung ventures less profitable than they might otherwise have been. This change arose from the development of the freight-car cooled by ice, and then the application of this concept to the railways. It was at this stage that the use of low temperatures really started to change the face of society and to alter the way of life of large numbers of people.

This change is best exemplified in the growth of the city of New York. In the late 1830s there were some dairy farmers with regular runs of more than a hundred miles into the city, carrying butter and other perishables in ice-cooled freight cars. At the same time the use of ice for food preservation allowed the establishment of greengrocery, produce and butchers' shops in outlying parts of the city. No longer was it necessary to go into the central markets daily in order to obtain fresh food. Old ordinances about the control of these central markets then fell into disuse and the city could spread horizontally over the surrounding countryside. In August 1855 a correspondent of *Hunt's Merchants' Magazine* was reporting that 'In summer no fresh provisions are on display at the markets at 10 a.m. but every variety is to be found in hundreds of ice chests in which they are stored'. In this same year the famous Knickerbocker Ice Company was founded and ten years later the ice-dealers of New York sold 104,500 tons and had a further 57,500 tons of ice in stock. To cope with this sort of demand the men of Massachusetts were harvesting more than 650,000 tons of ice, valued at more than 735,000 dollars in their best year. Special light railways were built to connect the best ponds to the ports, and Wyeth constructed a brick-built ice house of 36,000 square feet, capable of holding 40,000 tons of ice.

The great saga of ice on the West Coast of the USA began in 1850 as a direct consequence of the California Gold Rush of 1848. There is virtually no natural freezing on the West Coast south of what is now the Canadian border. At the time of the Gold Rush, Russia owned Alaska and the British Hudson Bay Company controlled Oregon in the interests of fur-trapping even if it did not formally 'own' that area. To satisfy the luxury demands of those who had 'struck it rich' in California, the barque *Zingari* landed 50 'refrigerators' (ice-boxes) and 275 tons of ice from Boston in San Francisco and Sacramento in the summer of 1850. An ice delivery service with wagons labelled 'Boston Ice' was soon in operation and shortly afterwards the ship *Lucas* established further trade. She landed ice, and apples stored in that ice, and then she sailed northwards to try to buy more ice from the Russians on the Alaska coast. This fell through since the Russians asked too high a price for the commodity.

The American-Russian Commercial Company was formed in 1853, however, and from its San Francisco base it started trade in ice from the Aleutian Islands. The natives were trained to harvest the ice and a lumber mill was set up on the Russian-owned mainland coast with the principal object of supplying sawdust to insulate the ice in transit. At first, attempts were made to ship fish into California with the ice, but the trade eventually ceased because California had nothing to offer the Russians on the ships' return journeys. This one-way trade position put the price of ice up to 25 cents per pound weight. (Incidentally, in their native country the Russians had long practised a primitive form of ice-harvesting on the River Neva to supply the luxury trade of the royal capital of St Petersburg.)

Competition to Russian Alaskan ice came from the Nevada Ice Company, which harvested ice at Pilot Creek in the Sierra Nevada, and set up a depot in Sacramento. Soon this natural ice from the mountains was able to compete successfully with Boston ice for the markets of Hawaii and the Central American States.

In the early days of ice-cooled freight cars, some perishables and vegetables had been carried successfully to the Middle West from the then rich agricultural areas of New York and Philadelphia. But the serious development of long-distance refrigerated transport began in Ohio and the Middle West and coincided with the development of the railroads. Essentially, this was the start of the great meat-packing trade of the Middle West, but it received also a considerable fillip from the heavy German immigration of the 1860s. The Germans brought with them a taste for lager and the lager-brewers came too. They alone were using a million tons of ice a year by the end of the 1860s.

Before that, however, refrigerated rail freight cars were carrying Middle West beef and pork wherever the railroads went. John L. Schooley, in Cincinnati, modelled the entire packing room of his pork factory on a domestic refrigerator. Benjamin L. Nyce, of Decatur, went one further and used the same type of facilities for storing fruit. By 1857 refrigerator car shipments of meat and butter were arriving in New York from the West after spending nineteen days on the journey.

With the ending of the Civil War and the linking of East and West Coasts by the transcontinental railroads, shipping of refrigerated produce became nationwide. Special lines of freight cars in their own liveries, such as the Blue Line, were operating. The different railroads had to standardize their gauges largely at the behest of the refrigeration transport companies. Refrigerator cars were fairly simple—'ice boxes on wheels', as they were described —with ice stored in a central box and draining away through the floor as it melted. But there were many and rapid innovations in the technology, such as V-shaped troughs running the length of the car, and the use of brine instead of ordinary iced water. Fresh strawberries came from Chicago to New York in 1869, and three carloads of beef daily were going in the same direction by 1873. Ice-refrigerator cars started crossing the whole continent, being repacked with fresh ice as they crossed the mountain divides in Omaha, Utah and the Sierra Nevada. In the 1870s the East Coast sent eggs, grapes and peaches to California, while the West sent salmon back to the older States.

This vast trade in perishables, to and fro across the whole continent, was supported almost entirely by natural ice, harvested every winter from rivers, lakes and creeks, and stored throughout the summer months in huge warehouses. Because the ice was natural, and formed entirely free of charge, the development of ice-making machinery and the penetration of machines into the US market was held up. Leadership in the technology of refrigeration machinery fell to Britain, followed by France and Germany.

In 1896 the ice-harvest along the Hudson River and in Maine came to four million tons. The Knickerbocker company alone was employing fifteen to twenty thousand men, harvesting ice, using electric lights at night, and storing up to 90,000 tons in a single warehouse. Tugs pulled chains of ice-barges along the Hudson River. It was not until the end of the first decade of the present century that mechanical ice-making plants turned out more ice than was harvested naturally. Probably the crucial year was 1914, when the official estimates showed plant-ice production at 26 million tons and natural ice harvested at 24 million tons. At the same time the USA virtually ceased to export ice in any large quantities. The export market, started by the brash Bostonians like the Tudor family just after 1800, had reached a peak in 1870 when the USA sent out more than 65,000 tons of ice valued at more than 267,000 dollars. Boston was still responsible for nearly 90% of the ice exports in this decade and the largest portion of the ice still went to what the official statistics described as 'the British East Indies'. But by 1910 exports were down to a mere 23,000 dollars worth and the US was importing ice from Canada.

The tardy development of refrigerating-machine manufacture in the USA, in face of the natural bounty of lake and river ice, had profound effects on the international meat trade. In the 1870s trade in refrigerated goods across the Atlantic to Britain and Europe had expanded rapidly. A few cases of peaches and some poultry had actually been sent to London as early as 1849, when a writer in Chambers' *Edinburgh Journal* had welcomed them in the Free Trade spirit of the middle nineteenth century: 'If means could be contrived for transporting meat in ice at small cost, Europe would present a steady market for the surplus beef and mutton of America.' By 1879 America was sending 16,000 tons of butter a year to Europe and the principal importer was Britain who obtained 15% of her butter from the USA. But meat proved more difficult to manage, and it was not until 1875, the time of the rise of the Swift and Armour meat packing enterprises of Chicago, that any major development took place.

John Bate of Brooklyn had already made a name for himself as an inventor in the technology of refrigerated railway freight cars. In 1875 he fitted an insulated compartment in the British White Star liner *Baltic*, and he mounted an ice-bunker with a fan to circulate cold air between the decks of the ship. With this apparatus six quarters of beef and some mutton and pork were successfully transported to Britain and arrived in good condition. A rival system was immediately fitted into a liner of the rival Cunard line—this system had a solution of brine which passed through a coil of pipe in an ice box and then forced the cooled liquid through pipes in the meat compartment. In 1876 some 10,000 tons of meat was shipped in this way to Britain by a number of companies in New York and Philadelphia.

The equipment was, however, very bulky and carried comparatively small quantities of meat on each trip. In the 1880s, when the Chicago meat packers were sending more than 100,000 tons of dressed beef each year to the four largest cities of the American East Coast, British firms took the initiative and started fitting refrigeration machinery to ships. The Anchor Line vessel *Circassia* was equipped with a new system of refrigeration machinery based on the compression of air by a pump and its subsequent expansion to cool the meat compartment. This machinery was developed by the Bell Company of Glasgow with the aid of a technician, James Coleman, who used Lord Kelvin's principle (see Chapter 2). The Bell-Coleman machine was patented and had an important future. Once again the rival Cunard Line went into competition and installed a rival refrigeration system—the Hercules ammonia compression system—aboard several of its ships. The transatlantic trade in meat expanded until eventually the American meat-packing companies had their own stalls in Smithfield Market in London, the traditional centre of Britain's meat trade.

But this meat was only chilled, because the entire American refrigerated transport system depended on ice for cooling, and ice at that time was mostly natural, though some ice-making machines were proving successful. The real future of large-scale intercontinental meat trading lay with truly frozen meat, meat reduced to a much lower temperature than could be obtained by mere cooling with ice. The ships that were to carry frozen meat across the world would be equipped with refrigerating machinery so powerful that the freezing process itself would be carried out on board as the meat was being loaded.

The development of the frozen meat trade was to have enormous effects on the southern hemisphere countries of Australia, New Zealand, Argentina and Uruguay. At the same time the availability of vast quantities of meat from abroad was to turn Britain into a predominantly urban country, a nation which could barely support half its population from homegrown food, so completing the process started by the Industrial Revolution. Similar fundamental changes in society would later affect European countries such as Germany, France, Belgium and Holland.

It was quite clear to thinking men in Britain in the 1860s that some method of importing meat in large quantities was essential in order to feed the often hungry masses of workers in the new industrial cities. According to *The Times* in 1867: 'Never, it would seem, was there a clearer instance of the law by which one part of the world is fitted to supply the wants of another. But the wise provision of nature is defeated by the stupidity of man, and ten thousand sheep a week, instead of regenerating the poor of London are boiled down into tallow. . . . Such a story reads like a reproach to modern science.'

At about the same time the eminent divine, Dr Long, wrote: 'The wholesale and enormous destruction of valuable animal food going on in New South Wales for eight years. . . . Viewed in connection with the fact that there are millions 'at home' on the brink of starvation . . . is discreditable to Great Britain and her rulers and cannot but be peculiarly offensive in the sight of Heaven.' With this climate of opinion prevailing, it was the Society of Arts in London that set up a Food Committee charged with the duty of discovering 'a means of importing meat, if possible in a fresh condition'.

The background to this situation was made up by the economic and social trends on opposite sides of the planet. In Britain there was the well-known growth of population and growth of cities that were features of the process we call the Industrial Revolution, while there was a strong political belief in Free Trade from the middle of the century onwards. What is less well known is that the national herds of livestock were steadily decreasing in numbers throughout the middle decades of the century. In 1860, for instance, the number of livestock animals in Britain fell by 4 million in a total of 46 million. Correspondingly prices rose, beef going up from an average of fourpence three-farthings per pound in 1851 to an average of eightpence farthing per pound in 1881. The forces involved were not simply economic. Britain had long used its physical advantage as an island to control the cattle plagues that were a feature of life on the continents. A policy of stamping out cattle disease by slaughter is first known to have been tried in 1348. Real progress was achieved from 1770 onwards by the total ban on importation of live cattle and products such as hides. While Europe in the Napoleonic era was ravaged by rinderpest, pleuro-pneumonia and foot and mouth disease, Britain remained free of these cattle diseases, of which rinderpest, endemic in Russia, was probably the worst. With the coming of Free Trade after the repeal of the Corn Laws in 1846, the regulations against import of cattle were also slackened. As a result, in 1865 the import of a cargo of 331 live cattle from Revel in the Baltic to Hull brought rinderpest again. More than 73,000 animals were affected in the ensuing plague. A slaughter policy was necessary and the ban on the import of live animals was to some extent renewed. It was permitted, however, to import live animals provided they were slaughtered immediately on arrival, thus allowing the sale of meat, bred and raised in countries such as the USA, but only in an inefficient and necessarily small-scale operation.

In Australia and New Zealand, however, the flocks and herds had been growing apace as more and more country was opened up. Wool was the product that turned Australia from a struggling, half-penal colony into a rich country moving towards independent nationhood. But there were large beef herds there too. New Zealand, physically smaller, was even better adapted to the raising of high quality sheep, though not so successful with cattle. New Zealand did not suffer from the droughts that afflicted Australia, nor from distractions like gold-rushes, and in the long run was to provide rather more of Britain's food than the larger country. Both countries produced dairy products. Argentina was in a roughly similar position, though for historical reasons that country had concentrated on the production of hides for the Spanish leather industry rather than wool.

These countries, therefore, had huge herds of livestock raised only for the value of their outer coverings. It was equally clear to the inhabitants of these developing nations that they needed some way of transporting the meat to Europe. The only product of the carcasses they could sell was the tallow (fat), made from boiling down thousands of bodies.

Various methods of exporting meat were tried. The most nearly successful was the trade in canned meat from Australia. An exhibit of this product was a feature of the Great Exhibition of 1851, and a particularly determined effort was made to sell canned meat to the armies in the Franco–Prussian War of 1870. Large tins containing thirty pounds of meat were despatched but the contents of many were found in 'an advanced state of decomposition'. Not all the attempts were technical failures and quite large amounts of tinned meat came on to the English market, but this was not appreciated 'by those classes to whose use it is specially adapted'—in other words, even the poor would not buy it.

Another avenue explored was the making of meat extract by the process invented by Liebig, the then famous Austrian scientist. And more than two hundred patents were filed in Britain for various processes of preserving meat. They included coating the meat in tallow, dipping it in bisulphide of lime, and even killing the animals by making them inhale 'carbonic acid gas'. There were no major successes. It was also quite clear that no supply of ice could be made to last the journey from the other side of the world and therefore there could be no application of the American method of sending beef chilled by the use of ice.

This negative point was proved by James Harrison, possibly the greatest pioneer of the frozen meat trade, an inventor of genius, yet a man who failed in almost every business venture he tried. His tombstone records, 'One soweth—another reapeth'.

Harrison was born in 1815 in Dumbartonshire, was apprenticed in Glasgow and studied chemistry at Anderson's College in that city. This biographical information is highly relevant to the general story of low-temperature science and technology, for a great deal of the earliest work on refrigerating machinery and much research on ice and ice-making was performed in Glasgow. At least one modern authority suggests that the sight of snow on the hills with steam from the distilleries in the valleys was an incentive to curiosity and experiment. Harrison may well have imbibed some of this native wisdom before he emigrated in 1837 to Australia, where he started to prosper as a journalist, and became a member of the Legislative Council of the new state of Victoria.

In the 1850s he started to experiment with ice-making and built a primitive machine based on the evaporation of ether. His trials were a failure—for which he blamed the 'inferiority of Colonial workmanship'. But Harrison had been corresponding with Michael Faraday, then at the height of his career at the Royal Institution in London, a pioneer of low-temperature science and the liquefaction of gases. So he came back to Britain and persuaded Dr Daniel Siebe, of Denmark Street, Soho, to make a machine to his specifications. The more sophisticated workmanship available in London made the machine a success, and it was exhibited to the public in 1858 making three tons of ice a day.

(The American-born Jacob Perkins had in fact invented the cycle of evaporation, compression and condensation of a liquid for refrigeration purposes and patented it in London in 1834, but he had never made more than a working model, and his system was not mentioned in print till 1882, although it is the basic process used in the majority of refrigeration machines today.)

After his triumphant demonstration of his method, Harrison took another machine manufactured by Siebe back with him to Australia, while Siebe, after making the world's first refrigerating machine to go into regular production, continued marketing the design and thus founded the modern refrigeration engineering industry.

Back in Australia, Harrison demonstrated his plant, showing carcasses of beef, fish, poultry and sheep frozen by his ice. At the Melbourne Exhibition of 1873 he showed beef and mutton that had been kept successfully for 85 days. This not only shattered the current belief that the 'life' of frozen meat was only 19 to 21 days, but was long enough to cover the length of most voyages to England. It gained him enough financial support to start converting one of the holds of the sailing ship *Norfolk* to carry fifteen tons of meat at 25°F in tanks surrounded by a mixture of ice and salt. On 23 July 1873, the *Norfolk* sailed from Sandridge Pier. But the work had been done too hurriedly while 'the carpenters were not quite sober', as Harrison later told his supporters. Only a week out at sea, a fatal flaw in the scheme showed up when liquid brine began to leak into the ship's bilges. Nothing could be done to repair the failure and by the time the ship reached the Azores the meat had to be thrown overboard.

No news of this reached London in advance of the *Norfolk*, so the members of the Food Committee of the Society of Arts went down to Gravesend to meet the ship when she arrived on Sunday 19 October 1873. They had a wasted journey and Harrison had to tell them that he had to 'describe a success and record a mischance'. It became clear within the next couple of years that refrigerating machinery would have to be installed on the ships themselves and not just used to prepare ice before the journey. Harrison, who had blazed the trail and set men's thoughts moving in the right direction, returned to Australia to end his life in failure.

A similar fate awaited the efforts of the next pioneer, Thomas

Sutcliffe Mort. He too was born in an area of Britain much affected by the processes of industrialization, in fact at Bolton, in Lancashire, and emigrated to Australia, where he made a substantial fortune in the wool business. Faced with the wastefulness of boiling down his sheep for tallow, after they had given up their wool to the shearers, he started experiments on meat preservation, working with the French engineer, Nicolle. Mort rejected the idea of cooling the meat with ice or brine right from the start. In 1861 at Dorling harbour, Sydney, he built the world's first freezing works, in which animal carcasses were reduced to temperatures below freezing by direct refrigeration. Mort used the evaporation, compression and condensation cycle of a liquid to achieve refrigeration, but he used ammonia as his working liquid. Not until 1875 did Mort feel able to make a really large public demonstration of his success. But then he arranged a giant dinner party and fed his guests with beef frozen in the previous year. In his speech he proclaimed: 'I now say that the time has arrived—at all events is not far distant—when the various portions of the Earth will each give forth their products for the use of each and of all; that the over-abundance of one country will make up the deficiency of another; the superabundance of the year of plenty serving for the scant harvest of its successor; for cold arrests all change. Science has drawn aside the veil and the plan stands revealed. Faraday's magic hand gave the keynote and invention has done the rest.'

Mort showed, in this purple passage, that he had a clear understanding of what he was trying to do, for he understood that 'cold arrests all change'. Most others at this time were seeking solutions based on what they believed was 'the antiseptic power of ice'.

In the following year, 1876, Mort chartered the ship *Northam* and fitted her with an ammonia refrigerating machine, which cooled liquid brine; this in turn circulated through pipes inside the insulated compartment that was to contain the meat he proposed to ship to England. But in the loading of the carcasses, probably, these pipes were damaged. The motion of the ship, even in harbour, made things worse and the meat had to be unloaded before it could even leave the harbour. Mort had invested £80,000 in all his experiments, and he had persuaded others to subscribe £20,000 towards his first abortive attempt to ship meat abroad. His death in 1878 was hastened by 'the terrible blow' of his disappointment. His assistant, the Frenchman, Eugene Dominique Nicolle, remained at work, managing an ice-making factory which had to compete with natural ice still being shipped to Australia from America.

France, however, was at this time ahead of Britain in the development of refrigerating machines, and once again it was the appetite for ices and iced drinks which forced the pace. According to one authority, 'By 1850 the demand for cool drinks in the cafés and restaurants of Paris had increased beyond the tedium of perpetual hand preparation of freezing mixtures. . . .' The two brothers Carré, Edmond and Ferdinand, developed different kinds of refrigerating machinery. Ferdinand worked on the same lines as Perkins and Harrison, using ether in a cycle of compression and evaporation. Edmond, however, produced a device which refrigerated by absorption of a liquid in a vacuum—based on ideas produced by another Scot, John Leslie of Edinburgh University.

This type of absorption refrigerator, though using ammonia as the working liquid, was pursued by another Frenchman, Charles Tellier. But Tellier went back to the ammonia evaporation and compression design when he fitted refrigerating machinery in the ship *City of Rio de Janeiro* and transported 300 kilos of beef *from* London to Montevideo. It is still not clear why Tellier should have wished to perform this journey, but 23 days out from port there was some accident to the machinery and the meat had to be eaten on board.

That failure was in 1868, and it was not until 1875 that Tellier made his next attempt, supported by a company formed in France with the object of importing meat from Texas or Madagascar. A slow steamer, the *Eboe*, 210 feet long, was purchased and transformed into the *Frigorifique*, equipped with three Tellier machines. These cooled three meat-holds, each 85 feet long and 25 feet broad, insulated with cork and chaff. Again Tellier took meat from Europe when he sailed from Rouen in September 1876 and arrived at Buenos Aires on 5 December.

When the meat was taken out of the holds 'dark spots' were found on it. Nevertheless it was served to assembled official guests. 'At table they gave us small dishes prepared from the meat, the flavour of the most part of it was rather unpleasant.' It received what might be called a cool reception. Perhaps because of this,

Tellier had difficulty in assembling a return cargo. Nevertheless, on 14 August 1877 the *Frigorifique* docked at Rouen after 104 days at sea. By this time some of the meat had been frozen for 110 days and parts of the cargo were not in good condition. More friendly commentators were now to be heard, who merely noted that 'a rather careful selection had to be made'. Tellier's feat achieved a good deal of publicity and a Paris paper (*Le Rappel de Paris*, 2 December 1877) claimed that 'the problem is solved'.

Technically, indeed, Tellier had solved the problem of long-distance transport of meat, but his effort was an economic and business failure. The French buying public would have none of his frozen meat. The ship itself was in trouble, nearly knocking down one of the Seine bridges. He even had to take ten tons of his meat to the London market in an effort to sell it. It aroused much technical interest at Smithfield—'not an atom of mould was on it'—but it was 'like leather' and had lost 30% of its weight.

'The first entirely successful frozen meat shipment in the world's history' was Tellier's achievement within a year of the half-success of the *Frigorifique*. S.S. *Paraguay*, fitted with Tellier refrigerators, started loading meat and freezing the carcasses in her own hold at Buenos Aires on 7 October 1877. She had a collision on her journey and had to put into St Vincent's in the West Indies for four months for repairs. Nevertheless, when she arrived at le Havre on 7 May 1878, all the 5,500 mutton carcasses were found to be in perfect shape. The reporters at le Havre were enthusiastic; 80 tons of meat were used 'to the last morsel'; the garrison troops feasted on the mutton—'The congealing completely destroys the germ of putrefaction.' Even the Grand Hotel in Paris served the Argentine meat for a week. But Tellier and his frozen meat were abandoned by the French housewife and chef, and the *Paraguay* sailed no more.

For an Englishman, brought up to believe that the national failing is the inability to turn to profit the results of scientific laboratories, the most surprising remark in the whole history of the low-temperature industry comes from an expatriate Frenchman, Dr Pierre Berges, of the National Bacteriological Institute of Buenos Aires, who became the unofficial chronicler of the Argentine meat trade. He wrote, just after the beginning of the twentieth century: 'As has often happened in the history of industries, it has been the French who have made the discoveries and the

English who have turned them to account to their profit. The refrigerating industry belongs to this number.'

The English may, eventually, have taken the profit, but it was the Australians who took the initiative. News of the *Paraguay*'s performance reached Queensland in early 1878, and a group of squatters (nowadays we should call them cattle ranchers) got in touch with their business contacts and agents in London and sent them off to inspect the *Paraguay* at le Havre. Mr Andrew McIlwraith and Mr Beardmore Buchanan reported against the idea of using Tellier's Carré-type absorption machines for Australian conditions. But, bearing Mort's failure in mind, while at the same time accepting his theories about the necessity of freezing the meat rather than chilling it, they approached the Bell-Coleman Company of Glasgow. This was the company that had pioneered the shipping equipment for American chilled beef. Bell-Coleman were asked to provide machinery which would supply cold air to a ship's hold so that meat could be frozen as it was loaded and kept fresh throughout the voyage from Australia to London. The McIlwraith–McEacharn Company took the financial risk and chartered the S.S. *Strathleven* for conversion to refrigerated carriage. The operation was carried out with some thoroughness. There were experiments with the new equipment at the Bell-Coleman works. A civil engineer, Mr James Campbell, was put in control of the machinery for the voyage. Representatives of Bell-Coleman and of the Queensland cattle men, and the man who was later to be Sir Thomas McIlwraith, one of the giants of the frozen meat trade, all went along.

The *Strathleven* sailed out from England in the summer of 1879 and was loaded with fresh produce, forty tons of beef and mutton, at Sydney and Melbourne, freezing the cargo with her own machines as it was loaded. She set sail from Australia on 6 December 1879 and arrived in London on 2 February 1880. 'On inspection of the meat while the vessel lay in dock it was found to be in a perfectly sound state, frozen quite hard and covered with an artificial rime.' This report from the New Zealand product representative in London was quite possibly the most influential of all, because it caused a stir of activity in his own country which was to bring important results.

The frozen meat received a good press. The *Daily Telegraph* reported, 'It has been tested by the ordinary methods of cooking and found to be in such good condition that neither by its appearance in the butchers' shops, nor by any peculiarity of flavour when cooked for the table, could it be distinguished from freshly killed English meat.' A carcass of lamb was sent to Queen Victoria, a sheep to the Prince of Wales, a joint was presented to the Travellers' Club, where Lord Hatherton reported his enjoyment and surprise. Perhaps most important, meat which cost a penny-halfpenny to twopence a pound in Australia realized fourpence halfpenny to sixpence a pound in London. The frozen meat trade was not only desirable at both ends of the chain, it was also going to be profitable.

A meat-freezing works was set up in Australia in that same year, but the dramatic adventure that caught public imagination was the arrival in London docks of the second Australian frozen shipment aboard the *Protos*, which landed on 17 January 1881. She brought 4,600 carcasses of mutton and lamb and 100 tons of butter. It was all in perfect condition and London was cut off from the country, at that moment, by a prolonged snowstorm. The meat fetched up to seven pence a pound and the butter one shilling and threepence a pound. 'A substantial profit' was made, and before the end of the 1880s the regular steamship lines, such as the Orient Line, had joined in the enterprise and fitted refrigerating machinery to their vessels. Successful pioneers of shipborne refrigerating machinery were the Haslam company of Derby.

The founder of this company ended life as Sir Alfred Seale Haslam, M.P., but he began as an apprentice in the Midland Railway works at Derby. He founded his own engineering company in 1868 and produced a 'dry air' refrigerating machine in 1880. This was exactly what the shipping lines needed, and the Haslam company had a virtual monopoly of the early machines for ships. Haslams were the first 'big name' in the refrigeration industry.

New Zealand had 233,043 sheep in 1851. Thirty years later there were 11½ million sheep on the two islands, owned by 6,857 farmers. Such was the wool boom, and such was the meat that could be taken to England when the transport problem was solved. W. S. Davidson was the New Zealand manager for a Scottish-based land company when news of the Australian success with frozen meat arrived. He persuaded his directors to authorize the expenditure of £1,000 to see if New Zealand could follow suit, and the Bell-Coleman company was set to installing one of its 'compressed air' refrigerating machines aboard the sailing ship *Dunedin*. Davidson was assured by the Glasgow engineers that the joints would be all right for a hundred days at sea, and he had meanwhile ordered the erection of killing sheds in New Zealand, the collection of a gang of the best butchers and the selection of the most suitable sheep.

Davidson himself supervised the first loading of the *Dunedin* at Port Chalmers on 7 December 1881. The sheep were killed and butchered ashore and then frozen aboard the ship. The chief question for Davidson was 'Whether the carcasses, after they had been frozen aboard the ship, should be placed "thwart ship" or "fore and aft" in the chambers.' However, after four days the crankshaft of the refrigerating machine engine broke because of a flaw in the casting. The 641 sheep already frozen and stored aboard and the 360 killed but not yet frozen had to be unloaded and sold locally. The first customers for New Zealand frozen lamb were New Zealanders.

Repairs were done, loading was started again and completed on 11 February 1882, when the *Dunedin* sailed for London. She took ninety-eight days over the voyage, arriving on 24 May, but the refrigerating machinery had worked steadily and had often been needed for only two or three hours in each day. There was more to the voyage than that, however. Davidson remembered the arrival in London thus: 'Captain Whitson came to London ahead of his ship in a pilot boat, looking very strained and care-worn as he entered the shipping company's office. He was not quite sure about the condition of the cargo but thought that most of it was sound. The vicissitudes of the experimental voyage were related, the captain's anxieties about the cargo having been aggravated by his dread that his masts would be burnt, as the sparks from the funnel set fire to the sails on several occasions. (This would be the funnel for the engine of the refrigerating machine.) When in the tropics the ship was for a long time on one tack and, owing to its steadiness, the cold air was not sufficiently diffused among the carcasses, and in fact the temperature in the upper chamber remained so high that the engineer was almost in despair.' Captain Whitson was obviously the right man for this voyage, however. To alter this faulty circulation of cold air, the captain himself crawled down the main cold-air trunking shaft to cut extra holes so that the cold air could get into the storage chambers. In doing this he became so numb that he could no longer move. The mate had to crawl in after him and secure a rope round the captain's legs and have him pulled out.

After these heroic efforts it was no more than justice that the 5,000 carcasses of sheep and lamb were found to be perfectly sound and were sold in Smithfield market within two weeks. Contemporary comment was: 'At first salesmen were rather doubtful about the venture being a success, especially as it was the first trial from New Zealand, but when they saw the fine big sheep, which, though many of them had been frozen for over four months, were as clean and bright as freshly killed mutton, they quickly changed their opinion and pronounced the meat to be as perfect as meat could be.' The achievement was even mentioned in the House of Lords.

Fortunately the exact accounts of the voyage down to the last half-penny have been preserved. Only one carcass was condemned. Ten were given away as presents. The average weight of each carcass was 81lb worth £1. 1*s*. 8$\frac{3}{4}$*d*. on the market. The weight loss from freezing was about one pound per carcass. Allowing a little for the value of the hides and tallow left behind in New Zealand, the profit on the voyage was £4,216. 11*s*. 11*d*.

Once the point was proved, the trade rapidly expanded. Freezing works were built in New Zealand like those in Australia. Argentina followed suit, building 'frigorificos'. Refrigerated stores were built in Britain to keep the frozen meat. Commercial rivalry developed among importers, exporters, shipping companies and UK butchers. The Australians and New Zealanders bitterly criticized the Smithfield market system, and later they opened up direct trade with other British ports such as Bristol and Liverpool. The frozen meat trade led directly to the setting up of the big retail butchery chains, which were followed in turn by chains of grocery shops and finally by the supermarkets of today. The South American trade developed on slightly different lines, with British-based firms setting up ranches and freezing factories in Argentina and Uruguay so that a single company controlled the entire length of the trade from animal on the hoof to butcher's shop. Australia had the misfortune to suffer a great drought for several years at the start of the 1890s and lost out on some of the

British trade. But she developed a fairly large trade with other parts of the world, notably the Far Eastern countries.

Nevertheless, by 1919 Australia was sending Britain more than four million sheep and lamb carcasses and more than half a million quarters of beef annually. New Zealand was sending more than six million sheep and lamb carcasses, and South America was sending three million quarters of beef. British engineering and shipping dominated the transport side of the world frozen meat trade, and had a large share of the world refrigerating industry. British finance and insurance established a dominance in international commodity trading. Refrigerating machines and ice-making plants were making inroads in the USA but natural ice still dominated. Meanwhile Continental Europe fought hard to keep frozen meat out.

The countries of Western Europe were less intensively industrialized than Britain at the opening of the present century and they were by no means devoted to the principles of Free Trade. There were politically strong 'agrarian interests' in most of them, and early attempts to send in frozen meat, often from the London market, antagonized the butchers' interest and the agrarians'. France was quite capable of feeding herself in 1880, and the government there not only passed a 'lung law', insisting that all imported meat should have the lungs included, but imposed heavy duties. These were raised further in 1891 to 33 centimes a kilo plus an 'octroi' of 12 centimes, so that Australian frozen meat was paying twopence per pound duty, or 43%. Italian duties amounted to 36%, Germany exacted 62% duties, and the Belgians had an even sterner 'lung law' than the French.

Germany had the strongest case for needing frozen meat; the country was heavily industrialized, and could not feed herself. But the Agrarian Party was strong and in 1903 it forced through a Meat Inspection Law, 'more correctly termed a meat exclusion law,' and regulations keeping out US packed meat as well as Australian and New Zealand frozen meat. In 1905 there was a working-class outcry against the high price of food. The result was a further increase in tariffs as the Agrarians triumphed. In 1910 there were riots in Austria and France and the Vienna mob was yelling, 'Give us frozen meat.' In the years immediately before the First World War both Austrian and Italian governments were steadily relaxing their opposition to imports of meat.

To tackle this opposition the Premier Congrès International du Froid was held in Paris in 1908, followed two years later by a similar congress in Vienna. Likewise an International Congress of Meat Inspection was organized. American, British, Australian and New Zealand delegates joined to sponsor a motion: 'That this Congress expresses its opinion that in order to reduce the cost of living to the working classes and to promote international trade, regulations which hamper the introduction into any country of frozen or chilled produce and the storage, distribution or sale of such produce in any such country should be modified or abolished.' The Dutch delegate, however, thought that frozen meat was inadvisable and unhealthy both for people and cattle, and opposed the motion. And at the Vienna conference in 1910 the German representative declared, 'When frozen meat is thawed it does not look like meat but like a dirty wet rag.'

The First World War broke up these patterns. Tariff-ridden Europe was changed for good. The British merchant navy went heavily into refrigeration and switched to buying from North America because convoys from there were easier to protect. There was a last patriotic flourishing of the American natural ice harvest to save ammonia, the working fluid of a huge proportion of refrigerating machines, for use in ammunition and explosives manufacture; after the war the machines returned to favour.

The start of banana importing from the West Indies to Europe, using refrigerated ships, rescued the economy of many of the Caribbean islands as their traditional sugar crops failed to command world markets.

The period between the two World Wars was marked by three developments of major importance. In laboratories, research scientists began serious studies of the mechanisms involved in the freezing of cells and the damage caused by low temperatures. In the homes of the USA, the ice-box began to be replaced at ever increasing speed by the true domestic refrigerator. In Britain the first industrial manufacture of ice-cream was started and expanded. All these three developments were to lead to the completion of the 'cold chain' which today, fifty years later, links the farm and the market garden and the fishing boat directly to our tables.

Ice-cream, as has been seen, was a luxury dish in Europe in the seventeenth century. Towards the end of the nineteenth century, with the undoubted rise in the standard of living in the industrial cities, there came a sudden increase in the number of ice-cream vendors in cities in America and Western Europe. These were usually very small concerns, often run by one man or a single family, and in Europe they were very frequently Italians—there are families with Italian names who have lived in big Scottish industrial centres for four or five generations and who now want their own tartans. They made their ice-cream at home, freezing the mixture with a combination of ice and salt around the container. They traded their products from small carts drawn round the streets by horse or donkey, or even propelled by hand. At the turn of the century there were around two hundred ice-cream salesmen in London alone.

This was all changed in 1922 when Thomas Wall started the large-scale manufacture of ice-cream at a plant in the West London suburb of Acton. Thomas Wall's family had a long history in the meat trade, making and selling sausages and meat pies. In those days, before domestic refrigerators had appeared in Britain in any numbers, this meat trade slumped badly in the hot weather of the summer, and it was in an attempt to find some employment for his operatives and plant during the dog-days that Thomas Wall started looking at the possibilities of ice-cream. He sent a team to study the new American ideas of mass production (which were not, however, being applied to ice-cream making), and he found a marketing idea to go with them, pedal-tricycles carrying insulated containers and marked with the famous slogan, 'Stop Me and Buy One.'

The first of these tricycles was purchased in 1922 for £6. By 1939 there were 8,500 of them on British roads, operating from 136 depots. And Wall's were the world biggest ice-cream manufacturers, a position which the firm still holds, having expanded its operations into all the major countries of Western Europe and North America.

From a technical point of view this implied an enormous spread in the techniques of low-temperature engineering. Refrigerated transport now reached into every street of Britain. Low-temperature manufacturing techniques had to be discovered and perfected. Low-temperature storage could now be found in small local depots and was no longer confined to huge cold stores dealing with bulk meat and butter in a few large centres. It also called forth the development and production of solid carbon dioxide—'dry ice'—which was used to cool Wall's mobile sales-points. (The gas carbon dioxide changes directly into a solid when cooled to minus 78·5 degrees without passing through a liquid stage; the solid crystals, rather like snowflakes, can be compressed into a material which looks like compressed snow, and which evaporates directly into the harmless gas as it heats up.)

The Second World War naturally halted the steady increase in ice-cream sales. But once it was over the big boom in Europe came in the 1950s when Wall's sales in Britain trebled from £14 million in 1950 to £46 million less than ten years later. This almost certainly corresponded with the public acceptance of ice-cream as a table dessert instead of just something that the children could eat on a holiday outing. Certainly in the last twenty years the consumption of sponge and suet puddings has gone down while the consumption of ice-cream, both as a dessert and in general, has continued to go up.

The domestic refrigerator had hardly made any impact in British and European homes in the early 1950s; market penetration in the UK was only 8% in 1956. Today that market penetration has reached 75% and UK manufacturers sell more than a million machines a year. In Western Europe the home refrigerator has made even more impact than in Britain; in Germany, Italy, France and the Netherlands it has achieved a market penetration of more than 80%. But the true home of the refrigerator is in the USA, where it has achieved 98% market penetration. Before the Second World War these machines were virtually an American world monopoly; the very few that could be found in Britain were almost without exception of American origin. (In the late 1930s when refrigerators were already widespread in the USA, I remember seeing one for the first time in the home of a family friend who worked for a US company in London. As a small boy I was principally impressed by the capacity of the 'fridge' to make one's own ice-cream at home.)

The significance of the domestic refrigerator lay, however, in the completion of the 'cold chain' which it eventually brought about. The concept of the 'cold chain' seems to have been invented as early as 1912, long before the technology or the science was available to make it practical. Two men who wrote the first history of the frozen meat trade, J. T. Critchell and J. Raymond

(1912, revised edition 1969), produced the idea. The 'cold chain' conceives the whole history of any item of food, from the harvest, catch or slaughter to the consumer's table, in terms of a series of links, every link needing the appropriate low-temperature technology to keep the food cold, and therefore fresh, tasty and free from germs. The main links in this chain are : cooling or freezing immediately after harvest or slaughter; cold storage awaiting transport; refrigerated transport, processing or packing; holding at distribution centre; transport to retail outlet; retailing; storage in consumer's refrigerator. To keep the food in prime condition, all these depend on refrigeration. Today we add even further links, both in the domestic and the industrial settings, of re-freezing and storing after cooking. This is the cold chain and, speaking of it, one of the outstanding authorities on refrigeration in Britain today said, 'Some of these links may be absent or weak, but they should all be there if the food is to arrive at the consumer in the best condition.'

The first links of this chain were forged by the commercial necessity of providing some means of getting food such as meat and vegetables and dairy produce to distant markets. The arrival of the domestic refrigerator allowed further links to be forged at the consumer end of the chain. The first small packs of refrigerated food, prepared in sizes suitable to individual consumers, appeared on the markets, mostly in the USA in the early 1930s. It was not until our own era of the supermarket that this method of presenting and selling food developed into a really large-scale operation. In the meanwhile, the strategic pressures of the Second World War had forged many extra links in the chain, notably in the area of mass refrigerated storage of perishables and the building of cold stores which this demanded. These same strategic pressures also led to the development of techniques such as accelerated freeze drying (AFD), and to the broad scientific exploration of the techniques and problems of freezing and storing foods at much lower temperatures than those attained in the early days of mere refrigeration.

All around us now there is the continuation of this process and the application of what seems to be the general rule in the low-temperature field—the colder the better. The domestic refrigerator is being supplemented by the home freezer. The cold chain is becoming the freezing chain. The suburban housewife with a garden can even complete the entire freezing chain from harvest to table using her own equipment, and her own produce.

Scientific discovery of the basic processes by which ice, refrigeration, and low temperatures generally, preserve food has been going on in parallel with the technological and engineering development of practical methods for achieving this preservation. Scientists often believe that basic scientific work must precede practical developments, but most of the early advances in low-temperature technology show that this is not the case. Preservation by cold was observed in accidental, natural conditions; it was applied to the normal conditions of our living and then the scientists explained what was actually happening.

Food is spoiled (from the human point of view) by a variety of causes. The most important of these causes is attack by micro-organisms, such as bacteria, yeasts and moulds. Indeed, at normal temperatures there is a race between man and the many bacteria which live on the same food as man, to get to the food first. The bacteria have many advantages in the race, not least that they can render the food uneatable by man long before they have used it all up. The products of bacterial life and digestion are often highly poisonous to man, and the bacteria that cause cases of food poisoning in humans are simply the clear winners of that particular race. But many micro-organisms are highly susceptible to differences in temperature. They have evolved to grow and spread more efficiently in the range of body temperatures of warm-blooded animals. When the temperature is lowered to below 5°C they grow only very slowly. All food-poisoning bacteria and all the creatures we call 'germs' are of this type. But there are some micro-organisms which thrive best at 60°C and which hardly grow at all at 'normal' temperatures. And there are others—the psychrophilic bacteria—which grow best at around 20°C and which continue to grow quite well at freezing point. Not until the temperature gets down to minus 10°C does their growth cease, and even then they are not killed outright, but only undergo a slow death rate. These creatures cause the spoilage of chilled meat and iced fish. So once again it appears that for protection against micro-organisms, the colder the better. For whereas ordinary refrigeration to temperatures around freezing point preserves food by slowing or stopping the growth of bacteria, it is essential to freeze to low temperatures to kill the bacteria completely and thus secure long-term preservation.

Even when bacteria are controlled by cold, there can still be spoilage of food by the chemical reactions between the constituents of the food. Fresh fruits and vegetables are still alive when they are picked, they continue to breathe (more accurately, to respire) and to ripen. Fats become rancid, bread goes stale and fish deteriorates even when bacteria and micro-organisms are controlled. Many of these processes are caused by enzymes, the substances responsible for promoting chemical activity in all living bodies; and enzymes cannot be separated out from food substances. But low temperatures, as a general rule, cause chemical reactions to proceed more slowly, just as most reactions go faster when the temperature is high. Refrigeration therefore usually slows down biochemical spoilage even when it does not stop it. And yet again this effect is increased by even lower temperatures.

Such a simple matter as the loss of water also affects food for the worse. It causes the proteins to break up—the technical term is denaturation of protein. The food loses nutritional value and the texture is changed permanently. Water loss can be made worse by cold storage, so control of the humidity of the atmosphere in big cold stores and refrigerated depots is necessary. Complete freezing controls other, more subtle deterioration of food which comes from the redistribution of water within the food. An example of this is the movement of water out of the body of a loaf of bread into the crust, which always occurs after baking, making the body drier and the crust less crisp. This is stopped by freezing, which stabilizes the water in the various parts of all baked products.

At this point, however, we come to the essential difference between modern deep-freeze techniques and the procedures of ordinary refrigeration which have grown up through the first half of the twentieth century. (In present usage the word 'chilled' describes products refrigerated to around freezing point, 0°C. Technical advances have made it perfectly possible to transport meat from South America to Europe at these temperatures.. 'Freezing' or 'deep freezing' means refrigeration at temperatures of minus 20°C or below. But 'frozen foods' still tends to refer to small consumer-sized packages of food.)

In living bodies, animal and vegetable, there is water within the cells and other chemicals are in solution in this water. Likewise, there is a solution of chemicals in water between the cells. However, if a solution of chemicals in water (say salt and sugar in water) is reduced in temperature, ice crystals begin to form in the liquid somewhat below 0°C. These ice crystals then start to grow as more and more of the water freezes. As there is less liquid water left, the strength of the solution grows greater, i.e. the proportion of salt and sugar in the remaining liquid grows higher. This process continues until some particular temperature is reached—called the eutectic temperature and dependent on the nature of the solution, and the chemicals in it—when all the water becomes ice crystals, the other chemicals also crystallize and the whole mass becomes solid.

A similar process will occur when living material is cooled to 0°C and below. The growth of ice crystals both inside and outside the cells can then physically damage and disrupt the structures of the cells, such as the cell walls. But it is now believed that even more damage is caused by the very high concentrations of chemicals in the residual liquids inside and outside the cells. The highly concentrated chemicals can react with each other at greater rates than usual and the very high concentrations can drive substances into or out of the cells in unusual ways because of the extreme osmotic pressures that are set up. These processes do not matter greatly in the case of meats that are being frozen, although they will naturally make frozen meat different from fresh meat to the taste. But with more delicate fruit and vegetable cells they can damage the 'flesh' so severely during freezing that only a 'mush' is recovered when the product is thawed out for eating.

The answer to the problem is to freeze the food rapidly, and then to ensure that it is stored at a consistently low temperature. Rapid freezing produces smaller ice crystals, and when research was first begun on a large scale in the 1920s and 1930s, it was believed that the beneficial results were almost entirely from this cause. Nowadays, however, it is believed that rapid freezing achieves its good results more by shortening the time in which the reactions of the highly concentrated chemicals can take place. The current definition of 'fast freezing' is that the centre of the product should be cooled from 0°C to minus 5°C in less than two hours and the product should remain in the freezer until the warmest part of it has reached minus 15°C. The concept of rapid freezing can be taken too far, however, and it is not possible simply to immerse the product in liquid nitrogen or liquid oxygen at temperatures nearly 200 degrees below zero. Such treatment turns the product into a solid, brittle mass which will splinter into pieces if dropped.

Fish is notoriously and historically one of the most difficult foods to preserve. All over the world primitive societies discovered methods of drying fish in air so that the products of the seaside communities could be exported. In medieval Europe the trade in salted herrings from the North Sea was an important economic factor. In the last twenty years there has been a virtual revolution in the fishing industry with the introduction of the freezer-trawler or the freezer-factory ship, in which freezing equipment is taken out to the fishing grounds and the catch is frozen, either filleted or whole, right in what is the equivalent of the harvest field.

The movement towards this solution has been long and rather slow, each step forced on the industry as the boats have made ever longer journeys to ever more distant fishing grounds. The bacteria found naturally in fish are of the psychrophilic type, those that can live at low temperatures, and this is a simple result of evolution for the temperatures at which fish live are low compared with temperatures on land. In addition, fish have very powerful digestive enzymes in their guts, and these chemicals act rapidly to start digesting the flesh of the fish immediately after death. Fishermen have tried to overcome these forces by cooling the fish-holds with ice; packing their catch directly in ice: and various improvements on this, under the general title of 'superchilling'. In the British fishing fleet the first freezer-ships started to come into service in the 1950s—notably the three vessels *Fairtry I, II* and *III*. Freezer-trawlers such as *Northern Wave* and *Lord Nelson* followed and became commoner in the 1960s. Similar developments occurred in the fleets of many countries.

We have now reached the point where the rivalries between the many fishing fleets have virtually stripped the seas close to major land masses (the steady decrease in North Sea catches over the last twenty years is a clear example). The building of freezer-trawlers extends the range of a modern fishing fleet to almost any part of the world, at least in theory, and nations such as Russia have built up immense new fleets based on groups of trawlers working round a 'factory' or 'mother' ship which may also carry freezing equipment. In self-defence, many nations are now demanding, either unilaterally or through the International Law of the Sea Conference, an area stretching two hundred miles from their coasts in which the fishing rights will belong exclusively to the coastal nation. The problems raised are immense. On the one side is the plain necessity for conservation measures, and the rights of many poor and less developed countries. On the other are commercial rivalries, historic traditions, the right of free access to the High Seas, and the apparent injustice to those countries which have little or no coastline. The situation has not been created entirely by the technological arrival of freezing, for Britain and Iceland have engaged in a 'Cod War' over fishing grounds which are not so far from the markets as to demand freezing equipment on the trawlers. But the trends are clear enough for all parties to agree that some international law is necessary.

The application of the ideas of the 'cold chain' can thus be seen to be effecting social change and causing international problems in such a traditional area of man's activity as fishing. The first links of the cold chain undoubtedly caused international political change by affecting the development of whole nations such as Australia, New Zealand and Argentina. They likewise caused social change by improving the diet of the urban working classes of western Europe and America. Now the forging of the later links of the chain are affecting farming and agricultural patterns. In order to process vegetables and fruit by refrigeration or freezing, it is essential to apply factory methods to farm products, and, in order to achieve the best results, the freezing must take place as soon as possible after harvest. Thus the farmer and the food-processor must plan the harvesting together, so that the crop may be immediately transported to the rural factory where the freezing equipment must be made to work at an economic optimum. The timing of the harvest may depend on the availability of freezing equipment just as much as on the weather. For the farmer, this means in turn that planting and cultivation will also depend more on the timing of factory availability than upon the optimum weather. Similarly, it forces groups of traditionally independent farmers into cooperative action to achieve the harvest in the minimum time. And it is one of the factors forcing ever-increasing mechanization upon the countryside.

The steady strengthening of the cold chain is, in fact, forcing a certain uniformity upon food production techniques, whether meat, fish, vegetable or fruit production. It is at the same time bringing a uniformity to patterns of urban life—strengthening the viability of city life by making it independent of the vagaries of weather and climate that formerly affected day-by-day transport of rural products into the towns. The widespread distribution of mass-produced, brand-name frozen foods brings uniformity to all our kitchens and dining rooms. And the pattern of supermarket and frozen-food cabinet is being imposed on every city in the world.

Surprisingly, in view of this uniformity being imposed on producer and consumer, the industries which supply the machinery of the cold chain have become, if anything, more diversified. The first refrigeration machinery was provided by the engineering industries of Britain and Europe. Later, American engineering joined in. And though commercial fortunes have varied, the supply and manufacture of refrigerating machinery and the provision of cold stores and refrigerated transport have remained part of the general engineering industry, with many companies involved, and no particularly dominant firms emerging. But our domestic refrigerators and deep-freezes, which complete the other end of the cold chain, have not, in the main, come from the 'heavy' end of the engineering industry. They are mostly the products of companies which specialize in consumer capital goods; that is to say, companies which also manufacture cookers and washing machines and electric heaters. The big food processers are independent of both these groups, and the majority of their well-known names have arisen historically from companies which have their origins in the food trades.

Now an entirely new set of forces are coming into this picture —they are the industrial gas companies, which have their origins in the chemical industry. The principle 'the colder the better' has combined with the need to find a market for a surplus by-product, to bring into the food-freezing business such giant companies as the Air Products Ltd and Union Carbide in the USA, L'Air Liquide in France and the British Oxygen Company. On the one hand, scientific research into the preservation of living material such as bull semen and blood showed that freezing and storage with liquid nitrogen at a temperature of minus 196°C was highly effective and economical. On the other hand, the demands of the steel and chemical industries for liquid oxygen left the industrial gas companies with large quantities of liquid nitrogen on their hands.

The word 'cryogenics' has been coined to cover both the science and technology of working at extremely low temperatures such as that of liquid nitrogen. The most exciting development in food technology of the present decade is the application of cryogenics to the cold chain.

Normal refrigeration by machinery is achieved by extracting heat, or energy, by some mechanical process. Because of its greater power and its flexibility in application, mechanical refrigeration ousted the traditional method of cooling by placing the object to be cooled in contact with a cold substance like ice. The present upsurge in cryogenic techniques reverts to the use of a very cold substance, and usually this is liquid nitrogen. All the big industrial gas companies now offer a range of devices for freezing foods at various speeds and in varying quantities by the use of liquid nitrogen. Foods so treated range from chickens and other poultry to cakes and confectionery covered with cream, including most forms of fruit and vegetables in between. And to follow up the freezing, there are, of course, liquid nitrogen storage devices, tanks and chambers.

These liquid nitrogen freezing machines are not simple—as explained above, food products dunked into baths of liquid at minus 196°C become brittle, tasteless lumps. The machines, which invariably have colourful names—Zip-Freeze, CryoQuick, Snow Spiral and Cryosas are just examples—are usually based on a moving belt system of some sort, in which the food is steadily made colder as it proceeds through a tunnel or under a line of hoods. Some products are actually pressed between plates cooled by liquid nitrogen. But this technique could obviously not be applied either to raspberries or to cream cakes. More usual is a system where liquid nitrogen is sprayed into the cold end of the tunnel. The gas vapourizes and grows warmer as it proceeds towards the end at which the food is entering. The food correspondingly gets colder as it advances into ever colder layers of nitrogen. Nitrogen is inert and non-inflammable, and the power supplies needed at the factory are less than with the older forms of cooling such as air-blast freezing. Different machines applying the nitrogen at different speeds and temperatures can achieve different types of freezing—thus a machine specializing in 'crust freezing' may be desired for a particular product.

The great change which cryogenics is bringing to the food production industry, however, lies not so much at the producer's end of the cold chain, but at the caterer's end. In the industrial production of cream cakes, for instance, the cake mixture can be rapidly cooled by liquid nitrogen machines as soon as it comes out of the ovens, then the cream decorations can be applied immediately on to the cold, firm cake and the whole can be carried forward in the same process, on the same moving belt, to deep freezing for storage and later transport to the consumer via the frozen food cabinets of the local supermarket. In the USA one plant alone freezes three million hamburgers each day, while the largest liquid nitrogen freezing plant in the world is used to process pizzas; a huge production line produces the prepared food and then directly freezes every pizza solid ready for storage and transport.

The logical development of this line of food freezing has now taken place. This is the freezing of whole cooked meals after the cooking has been completed. The movement in this direction has followed lines parallel to most of the other developments in food production and preservation, but it has happened more rapidly. The growth of chains of hotels and the expansion of airline catering, with its special demands for minimum space, have hastened the process. The objective is to make the finest products of the best chefs available to large numbers of people in locations far removed from the chef's kitchen.

Probably the first major step in the freezing of cooked food was made in the hotel trade with the central production of special items such as sauces under the direction of the most skilled chef, and their subsequent freezing and distribution to other hotels in the chain. Freezing of dishes prepared and cooked in a central kitchen made rapid progress in the world of convention banquets and business dinners. The coming of very rapid freezing by liquid nitrogen has hastened this process and the techniques have invaded airline catering, as well. Here liquid nitrogen has special advantages because it is inert and non-inflammable and is therefore very attractive to aeroplane operators. L'Air Liquide now freezes and stores in liquid nitrogen nearly all the food for Air France long-range flights, so that one can eat an omelette over Tokyo that has been prepared and cooked by a genuine French chef in a French kitchen.

It is but a logical extension of this continuing process to come to the latest idea which is to set up a 'cook-freeze' organization and production facilities for any large organization with catering needs, or even as a commercial operation. The type of organization which is envisaged as requiring a cook-freeze system is one in which there are large numbers of people to be fed, but where central cooking has not proved popular—a group of hospitals, the schools of a town, or even a large unit in an army. The economics of central buying, storing, food preparation and cooking for organizations like these have long been obvious. The problem has been to transport the cooked food so that it still makes appetizing eating in the scattered smaller consuming units, such as individual schools or hospital wards. Transport in insulated containers, hay-boxes or bain-maries have proved so unsuccessful that a number of large central catering organizations have been broken up and less economic but more satisfactory local kitchens have been reinstated.

The cook-freeze system, however, claims to offer all the advantages of central cookery with the delivery of satisfactory and attractive meals to the consumers by freezing the cooked dishes with liquid nitrogen immediately they leave the cooker. No more than local thawing and reheating is then needed. The implications may horrify some readers and appal many cooks. The central cookery or kitchen must be regarded as a factory and planned as such, if the benefits of cook-freeze are to be obtained. Flow-lines and production methods must be used in everything from the peeling of potatoes to the serving out of individual portions.

There will be many who will wonder whether all of these 'advances' in food technology using low temperatures are necessarily 'progress'. The conservative food-lover (including the author) will continue to prefer an individual meal, individually cooked. Such a choice is not always available, however. There can be no doubt that the strength of the cold chain has steadily improved the standard of food available to the city dweller (and that covers the vast majority of our population) over the last hundred years. There may come a day when we will have to struggle to ensure that the chain does not bind us too tightly.

4 Cold Air

It was chemists, rather than physicists or engineers, who pioneered the way to cooling the air we breathe to such an extent that it becomes liquid and can be separated into all its many constituents. In this case, scientific theory preceded engineering practice, and ironically led to the biggest of all the industries that depend on producing low temperatures.

Eighteenth-century chemists had established the atomic theory of matter and the theory of elements, the concept that all matter is made up of only a limited number of basic substances, and that the essential units of each one of these substances are individual, indivisible atoms. All the differences of appearance and behaviour of the different substances stem from the differences between their constituent atoms. These fundamental concepts also involve the enormous intellectual leap of recognizing that one substance, one element, can appear in different guises to our senses—it can be either a solid, a liquid, or a gas. We nowadays call these the three different phases or states of matter.

The most obvious way of changing a substance to a different state was by applying heat—we melt a solid metal into a liquid, we boil a liquid like water to turn it into a vapour or gas. But during the late eighteenth century a Dutchman, M. van Marum, showed that it was possible to achieve change of state by pressure alone. He was performing experiments to test Boyle's law—the law that says that the pressure of a fixed volume of gas will change directly in proportion to the raising or lowering of temperature. He found that common ammonia gas will turn into liquid if it is lightly compressed at what were then fairly low temperatures. We now express this by saying that ammonia is a highly volatile liquid—and we use van Marum's discovery to operate some of our refrigerating systems.

As a result of these theories and experiments, the great French chemist Lavoisier wrote, before the end of the eighteenth century: 'If the earth were taken into a hotter region of the Solar system, say one in which the ambient temperature were higher than that of boiling water, all our liquids and even some metals would be transformed into a gaseous state and become part of the atmosphere. If, on the other hand, the Earth were taken into very cold regions, for instance, to those of Jupiter or Saturn, the water of our rivers and oceans would be changed into solid mountains. The air, or at least some of its constituents, would cease to remain an invisible gas and would turn into the liquid state. A transformation of this kind would thus produce new liquids of which we as yet have no idea.' These were indeed prophetic words. They set scientists off on the hunt to find ways of liquefying air and its component gases. This hunt in its turn led to the race towards absolute zero. And today we still accept Lavoisier's concept. Yet it was very nearly a hundred years before any scientist could show a practical example in the laboratory of 'liquid air'.

In the first half of the nineteenth century a fairly large number of scientists worked on liquefying gases by pressure and some succeeded, as van Marum had done, with a few materials that normally appear as gases. But very few experimented with both temperature and pressure and no one could liquefy air or its main components, oxygen or nitrogen, or even hydrogen, whatever pressure they applied.

A French scientist lowered cylinders of oxygen and nitrogen to depths of more than a mile in the ocean, subjecting them to a pressure of more than 200 atmospheres with no liquefaction. (One atmosphere, the standard air pressure at the surface of the Earth, is a pressure of about fourteen lb per square inch.) A Viennese doctor, Natterer, who took an amateur interest in high pressure devices, reached a pressure of 3,000 atmospheres by the middle of the century but did not manage to liquefy air. The idea grew that oxygen, hydrogen and nitrogen, at least, were 'permanent gases'.

The true course to the liquefaction of all gases was plotted out by three men over the first seventy years of the nineteenth century. Paradoxically, they all three studied primarily the heating of gases and liquids. Charles Cagniard de la Tour, who was in 1822 an official of the Ministry of the Interior in post-Napoleonic Paris, carried out experiments of a distinctly heroic nature, involving at first strong-walled gun-barrels and later, when he turned to glass tubes, many explosions. He decided to study the effects of heating alcohol in an enclosed space. His first problem was to find out what went on inside his gun-barrel, so he put a ball of quartz into liquid alcohol at various temperatures and found its rolling about made a different noise from that made when it was enclosed in a tube of air. Judging by this sound, he concluded that when alcohol was heated to a high enough temperature in his sealed gun-barrel it eventually all turned to the gas phase.

Trying to see what happened, he heated glass tubes which contained different amounts of liquid. Almost all the tubes containing as much as half their volume of liquid exploded. Cagniard de la Tour survived to declare that at certain pressures and temperatures he had seen all his liquid suddenly turn into gas. This was not ordinary boiling—he had discovered what we now call the 'critical point', the combination of pressure and temperature which is the salient feature of our present understanding of the relationship between the liquid-phase and the gas-phase of any substance.

At almost exactly the same time—in 1823—the young Michael Faraday, then an apprentice to Humphrey Davy at the Royal Institution in London, was heating a sealed glass tube containing chlorine compounds. He was being watched by a friend of Davy's, one Dr Paris, and they were both amazed to see small drops of an unpleasantly greasy liquid form at the unheated end of the tube. Dr Paris believed that there must have been impurities in the chlorine materials they were investigating, but a note arrived next morning from Faraday claiming that the drops were liquefied chlorine gas.

Faraday duly became Director of the Royal Institution in succession to Davy and extended his work in two important series of investigations in 1826 and in 1845. He had realized after his accidental success of 1823 that both temperature and pressure had played a part in liquefying the chlorine since the heating of the tube had increased the internal pressure but the liquid had formed at the cold end of the tube. In his later experiments he deliberately cooled one end of his tubes by immersing them in a cooling mixture while he applied heat to the other end of the tube containing various materials. He repeated his liquefaction of chlorine and liquefied a number of other gases as well.

The man who brought sense and order to this increasingly confused picture was Thomas Andrews at Queen's College, Belfast (now the University). In ten solid years of detailed work, from 1861 to 1869, he concentrated on the easily obtainable gas, carbon dioxide—CO_2. He steadily plotted the variations between pressure and volume of the gas at a certain temperature. Then he selected another temperature and repeated the whole process. First he showed, as von Marum had done when liquefying ammonia gas, that Boyle's law was not always true; that pressure multiplied by volume is not always the same. He, too, found the critical point that Cagniard de la Tour had found before him—indeed, Andrews considered his work to be the logical extension of the Frenchman's observations. Eventually he was able to draw up what amounts to a map of the states of carbon dioxide. This makes it clear that above a certain temperature—the critical temperature—the substance can never become a liquid, however much pressure is applied. Below this critical temperature, which is also the temperature at the critical point, carbon dioxide can either be a liquid, a gas and a liquid in equilibrium, or a gas, depending on the relationship between pressure and volume. (And our present-day knowledge adds an area in which carbon dioxide, or whatever substance we are studying, will be a solid.)

The lasting importance of Andrews's work is that similar 'maps' or graphs can be drawn for all materials, and they are all of the same general shape, though the precise temperature and pressure at which helium will turn into a liquid is very different from the temperature and pressure at which liquid iron vaporizes. For our immediate purpose, the work of Andrews showed why none of the attempts to liquefy oxygen and nitrogen by pressure had succeeded and pointed the way to the missing factor in gas liquefaction—low temperature. We know now that the 'critical temperatures' at which oxygen and nitrogen gas will liquefy are minus 118 degrees and minus 147 degrees. Unless we can reach temperatures as low as these, attempts to liquefy air are doomed to failure.

Two men, at least, appear to have spotted the importance of low temperatures at some stage of their work and they succeeded in liquefying oxygen by two entirely different methods in exactly the same month, December 1877. This is one of the outstanding examples of the phenomenon of 'independent and simultaneous' discovery which is found throughout the history of science and which seems so baffling. It is probably an entirely artificial 'phenomenon' created solely by the activity of historians. At many stages in the progress of science it is fairly obvious what the next move, the next step forward, should be. Of course it is of great professional importance to the individual scientist to be first with some new discovery—the idea of 'priority' is one of the basic elements in the structure of the scientific society. So at any one time there will probably be quite a number of scientists working towards the same new advance. The likelihood that at least two of them will get there closely enough in time to dispute the claim to priority is fairly high. The historian records the disputants to the claim and nothing is heard of the others who were not even among the 'placers' in the race.

However, there was a peculiar drama about the discovery of the liquefaction of oxygen. Rumours of the announcement of a great new discovery had circulated in Paris for some days before the Christmas Eve meeting of the Academie des Sciences. At this meeting a communication was read from one of the newly elected members, Louis Paul Cailletet, describing an experiment in which he claimed (quite accurately as it turned out) that he had observed liquid oxygen. The paper was immediately followed by an announcement from the Secretary, M. Dumas, that the Academie had received a telegram two days earlier, on 22 December, from Raoul Pictet, a physicist in Geneva, that he had liquefied oxygen 'at 320 atmospheres and 140 degrees of cold by combined use of sulfurous and carbonic acid'. An account of his method had been sent to the Academie earlier and the telegram was a claim (equally accurate) that the method had succeeded. Now priority is normally accorded to a scientist on the basis of the date of formal 'communication' to some appropriate learned body or journal. In France, at that time, it was communication to the Academie that really counted, so it seemed, according to the rules of the game, that Pictet had won by two days. However, there was more substance to Cailletet's position than at first appears.

Cailletet, the son of an iron founder at Châtillon-sur-Seine, was a professional mining engineer with a keen interest in science. He was standing for election as a corresponding member of the Academie des Sciences in the early weeks of December 1877. It was a contested election, with voting due at the 17 December weekly meeting of the Academie. Cailletet had actually performed his crucial experiment on 2 December, but he rightly judged that to announce such an important advance at any of the meetings just preceding the election would be considered bad form. So he had to wait till the 24 December meeting, even though he was successfully elected by 33 votes to 19 on 17 December. Very sensibly, he had taken the precaution of sending a letter describing his work to the Academie through a friend on 3 December. This was deposited with the Permanent Secretary who signed and dated it on arrival, allowing it to remain sealed, as requested. As a further precaution, Cailletet had given a successful demonstration of his oxygen liquefaction before an audience of friends and colleagues at the École Normale on Sunday 16 December. He was thus able to prove his priority over Pictet, though by a mere twenty days.

Cailletet had, in fact, been proceeding along the same hopeless road as many others, trying to liquefy gas by applying high pressures. He was following up a suggestion that it might be possible to liquefy acetylene (C_2H_2) at ordinary temperatures by applying pressures of around 60 atmospheres. He never got to his target pressure. The apparatus burst under the strain. But at the moment of the accident, as the pressure dropped inside the glass tube containing the acetylene, Cailletet saw a thin mist form, only to disappear immediately. He jumped to the conclusion that he had in fact seen the liquefaction of acetylene caused by the sudden cooling of the gas as the pressure dropped catastrophically. We now know that the cooling had been formed by adiabatic expansion, as explained in Chapter 2. It could, however, have been droplets of water forming the mist if there had been water as an impurity in his sample of acetylene. So he sent off to Paris to a famous laboratory for the purest sample of acetylene that could be made. With this he repeated his experiment, including the accidental drop of pressure. Again he saw the mist forming briefly, and he was convinced that his explanation was correct.

So Cailletet immediately set about repeating his experiment, accident included, on the most important subject, oxygen. It was not difficult, even in 1877, to prepare fairly pure samples of oxygen. He put his oxygen into a strong-walled glass tube which he cooled with evaporated sulphur dioxide. He probably got it down to about minus 29°C. Then he compressed the oxygen with a pressure of about 300 atmospheres. Suddenly he released the pressure and again a mist, a collection of tiny liquid droplets, appeared. Cailletet had only to perform some tests to convince himself that the droplets were not caused by some impurity and he could declare that he had liquefied oxygen. This was the experiment he repeated for his selected audience at the Ecole Normale. He well deserved his election as a corresponding member of the Academie des Sciences the following day, and he had justified his claim for priority for his announcement of 24 December.

The discoverer of liquid oxygen spent a very busy Christmas in 1877. Immediately after his triumph of 24 December at the Academie, he went back to work and, using the same technique, liquefied nitrogen. This result he was able to announce exactly one week after reporting the first liquefaction of oxygen. So the main constituents of the air we breathe had been turned into liquids at very low temperatures in the laboratory and Lavoisier's prediction had been fulfilled.

Ironically, it was the process adopted by the defeated Pictet that came nearest to the process used nowadays for liquefying air on a large industrial scale. Pictet used what we now call a cascade process to reach temperatures lower than anyone else had then attained. This uses a series of gases of progressively lower critical temperatures. The first, chosen because it can be liquefied by pressure alone, surrounds a vessel containing the second gas in the cascade. The first gas is liquefied and then allowed to evaporate. In doing so it draws latent heat from the second gas, cooling it in the process. If the system has been properly designed, this cooling is enough to bring the second gas below its critical temperature and liquefy it, although some additional pressure may be needed to complete the process. The second liquefied gas can then be used to cool a third gas which has an even lower critical temperature, and so on.

This concept explains the rather mysterious telegram Pictet sent to the Academie des Sciences, in which he claimed to have liquefied oxygen 'by the combined use of sulfurous and carbonic acid'. He said he had reached '140 degrees of cold' and had then had to apply 320 atmospheres of pressure to liquefy oxygen. It seems likely that Pictet's figures were overgenerous, for we know that the critical temperature of oxygen is only minus 118 degrees. Cailletet, whose method involved a virtual disappearance of pressure back to ordinary atmospheric level, calculated that his oxygen liquefied at a temperature of about minus 200 degrees. As far as we can calculate now, it seems likely that he was fairly accurate in his estimate.

Having liquefied both oxygen and nitrogen within a month Cailletet might have felt entitled to rest on his laurels. But at the crucial meeting of the Academie on Christmas Eve, 1877, one of the listening scientists (J. Jamin) had remarked, 'The decisive experiment has still to be made. It will consist of keeping liquid oxygen at the temperature of its boiling point.' In other words, he wanted to see a pool of liquid oxygen in a laboratory vessel, steaming gently. Cailletet spent many years working for this objective, but he failed. In 1882 he adopted half of Pictet's method, cooling his glass tube with liquid ethylene, so that the temperature was down to minus 105 degrees. But when he released the pressure on the oxygen within, there was no more than violent turbulence and a spray of droplets of liquid oxygen which disappeared immediately. Cailletet carried on working for the remainder of his life and he died proud to see a great industry developed from his experiment. His descendants still care for his laboratory, full of old instruments, at Châtillon-sur-Seine.

Cailletet had been one of those who believed in science 'pure and disinterested', a tradition which was handed on, particularly among French scientists, to the pioneers of modern physics and atomic physics such as the Curies and Rutherford. He published all his experimental details and encouraged the manufacture of large numbers of 'Cailletet apparatuses' for other scientists to use. One of these sets was purchased in 1882 by a Polish scientist, Szygmunt von Wroblewski, who had just been appointed professor of physics at the ancient and not very well equipped Jagiellonian University of Cracow. When the apparatus and von Wroblewski arrived back at the University at the start of 1883, they found a particular welcome from the young professor of chemistry, K. Olszewski, who was also interested in the liquefaction of gases. In two months the two Poles solved the problem of liquefying oxygen and keeping it liquid in the laboratory. They did this by very simple modifications to Cailletet's apparatus and it is quite probable that they succeeded where he had failed because they were used to struggling with antiquated apparatus such as universities obtained under the old Hapsburg Empire of

Austria. When able, then, to get their hands on some modern apparatus they were quick to improve it.

Like Cailletet in his later experiments, the Poles used liquid ethylene as their coolant. But whereas Cailletet had allowed his coolant to boil off at atmospheric pressure at a temperature of minus 105°C, von Wroblewski and Olszewski pumped off the vapour above the boiling liquid, thus reducing the pressure, which in turn reduced the temperature of the ethylene to minus 130°C, well below the critical temperature of oxygen liquefaction. (Basically the same process takes place when water boils at less than the usual 100 degrees where the atmospheric pressure is less than normal; for instance in Mexico City water boils at 93 degrees because of the reduced pressure so far above sea level.)

The Poles put their oxygen gas under high pressure in a steel cylinder. From the neck of this cylinder a strong-walled glass tube led into the bath of liquid ethylene. As a small final modification, the end of this tube was bent downwards so that any liquid oxygen produced would collect in the bottom rather than forming a mist, or running out through the expansion tap at the neck of the steel cylinder.

The two men started work in February 1883. Only two months later, on 9 April, they saw little droplets of liquid collecting at the cold end of their glass tube. These droplets came together and collected as a little pool of liquid oxygen which stayed, visible and satisfactory. They had not even had to reduce the pressure, in Cailletet's way, to reduce the temperature. The cooling provided by the liquid ethylene boiling at low pressure and the high pressure of the oxygen released from the storage cylinder had been enough to produce the desired result. Their success was communicated to the Academie des Sciences on 16 April 1883.

We still do not know which of the two men was responsible for the crucial new ideas, for their full paper, published later in *Annalen der Physik und Chemie*, was written jointly. But the tendency towards dramatic disagreement which marked so much of the early work on low temperatures was felt as much in Poland as anywhere else. Within a few months of their joint success the two men dissolved their collaboration and continued working separately on the same problems in the same university. Five years afterwards, von Wroblewski knocked over the oil lamp on his desk when he was working late at night and died in the subsequent blaze. Olszewski involved himself in furious public controversy with some of the most important figures of low-temperature science in later years.

The importance of the work of the two Poles was that they showed that liquefaction of oxygen could be more than a mere laboratory curiosity, more than a fleeting glimpse of a process which proved a theory. Their work was a starting point for three major developments.

First, there rose from the work at Cracow the start of the industry based on liquefying the gases of the air. Plainly, if liquid oxygen could be demonstrated boiling calmly in a test-tube in a laboratory, it could eventually be manufactured on an industrial scale. Cailletet had shown that nitrogen could be liquefied in his primitive way—by applying the Polish system it could be treated like oxygen, on a larger scale. And since oxygen and nitrogen are the main components of atmospheric air, it must be possible to liquefy air as well. It was on this theoretical basis that Carl von Linde and Georges Claude began the liquid air industry which will be more fully treated in the next chapter.

Secondly, the Cracow work demolished finally the idea of 'permanent gases' and led to the belief that maps and graphs of the type drawn by Andrews could be drawn for all materials. Thus it should be possible to study the nature of hydrogen and find methods of liquefying it, too. In the years ahead this also provided a rational approach to the liquefaction of helium. In broad terms, progress towards liquefying all gases could now be made on a logical, rather than a brute force, basis.

But to science in general the liquefaction of oxygen suddenly opened up the whole field of low-temperature studies—the area we now call cryogenics. Scientists had not considered the study of matter at temperatures below minus 100 degrees, simply because there was no way of attaining such states. Now the temperature could be reached in the laboratory and work could start in totally uncharted areas such as the behaviour of electricity and magnetism at low temperatures, the behaviour of living matter at this extreme, the effect of extreme cold on the brittleness of metals. The opportunities were limitless. A whole range of new sciences was born. And with their birth came an important shift of emphasis. No longer were low temperatures sought in order to liquefy new gases. Now the liquefaction of new gases became a means to achieve even lower temperatures. The race towards the unattainable goal of absolute zero had started.

It was indeed in the interests of science that the first air-liquefiers were made. The method pioneered in France by Georges Claude still works today in commercial applications and was a development of Cailletet's liquefier in that Claude relied primarily on the cooling produced by adiabatic expansion. But Linde in Germany chose to exploit the Joule-Thomson method of cooling which was described in Chapter 2. In fact William Hampson in England was working on the same method at the same time. Both Linde and Hampson realized that the cooling achieved in one passage of gas through a Joule-Thomson porous plug was very small, and both came up with the same answer: to weld together, side by side, the pipe that carried the gas towards the plug and the pipe that carries the stream of gas away from the plug. Thus the gas that comes away from the plug, the cool gas, will cool down the gas flowing into the plug. This regenerative system, though it works very slowly, will produce a major fall in the temperature of most gases if the system is carefully designed and operated.

Linde was the first to achieve successful liquefaction on a large scale. It was a heroic experiment that went on for three days, partly because heat leaked into the system each night, undoing much of the previous day's work. He wrote in his memoirs, 'We continued work for sufficient time to justify the expectation that a considerable amount of air had been liquefied. Then, amid rising clouds, we allowed the beautiful bluish liquid to stream out into a large pail.'

This was on 31 May 1895, and on 5 June 1895 Linde filed his patent. Hampson had filed his patent in England on 23 May of the same year.

The spotlight at the centre of the stage in the story of low temperatures now illuminates the figure of Sir James Dewar, Fullerian Professor of Chemistry at the Royal Institution in London. 'Spotlight' is not to be taken as a mere metaphor, for the most striking feature about the Royal Institution is its tradition of 'Friday Evening Discourses'—public demonstrations of the advances of science, given before a lay audience either by the director of the Institution himself or by a distinguished visiting scientist. These are not lectures, they are discourses, and an essential feature of them has always been the public demonstration of experiments which vividly illustrate the scientific points the speaker is making. The publication of science in this way is one of the objectives of the Royal Institution and enormous effort is put into preparing and perfecting the vivid experiments.

In high Victorian times, towards the turn of the century, these discourses attracted a truly distinguished audience of eminent scientists, politicians and other laymen, often accompanied by their wives, and dressed in the height of formal fashion—stiff shirts, tailcoats, full evening dress, with the ladies feathered and bejewelled. The tradition is carried on today, but it has changed with the times—many of the important series of discourses are now televised, and the experiments are designed to demonstrate the points to viewers at home as well as to the live audience in the famous lecture-theatre. The present director, Sir George Porter, and his predecessor, Sir Lawrence Bragg, both distinguished scientists, have become familiar figures on television.

Sir James Dewar was likewise a distinguished scientist in many fields. He is remembered now for his work in low-temperature science simply because that was the outstanding part of his research. And he simply loved the glitter of the Friday Evening Discourses with himself at the centre of the stage. This was but one of the characteristics which distinguish him from the stereotype of the eminent scientist. Another was his ability to construct violins, an art he had learned from the village joiner in his native Scotland, when he was confined to his home for several years by illnesses attributed to his having fallen through the ice into a pond. He also had a habit of strolling round the corridors and halls of the Royal Institution conducting conversations with the ghost of his distinguished predecessor, Michael Faraday—though this seems more understandable after one has been informed (as the author has been) by two distinguished living scientists that the presence of the spirit of Faraday and even his voice have been clearly noticed by them in the buildings in recent years.

The subject of the Royal Institution discourses is by no means irrelevant to the progress of low-temperature science. It was in order to perform experiments clearly visible to the public that Dewar invented vacuum insulation and perfected the 'Dewar flask'—our present day vacuum flask or thermos, the device which is still used in almost exactly the same form by industry and laboratories all over the world to store liquid gases.

As soon as Dewar heard about Cailletet's successful liquefaction of oxygen he ordered an example of his apparatus from Paris—he declared that he felt the spirit of Faraday commanded him to follow up this French work, which was itself a development of Faraday's own early experiments at the Royal Institution. By the summer of 1878 Dewar was demonstrating the mist of liquid oxygen droplets, in the manner of Cailletet, to his Friday Evening Discourse audience. But after that, he set about building up a more widely-based set of laboratory facilities for a major programme of low-temperature research in the old buildings in Albemarle Street, in one of the main shopping areas of London, between Piccadilly and Bond Street.

This took several years and it was a year after Wroblewski and Olszewski had demonstrated a continuous supply of liquid oxygen in their apparatus that Dewar was able to do the same. He had modified the Cracow apparatus somewhat, but mainly in order to let a large audience see clearly what was happening. He went on, however, to make further modifications to the experiment until in 1886 he was able to show the President of the Royal Society what solidified oxygen looked like. For some strange reason his first official description of the apparatus to achieve this was appended to a paper which dealt largely with meteorites.

Nevertheless, it was low-temperature work, and the development of the increasingly sophisticated laboratory apparatus necessary for this research, that mainly occupied Dewar's time. Much trouble was caused to all the early workers in the field by frost forming on the outside of their cold tubes and preventing them from seeing what was happening within. Cailletet, Olszewski and Dewar all developed various devices—now called cryostats—for containing their liquid gases. All had in common the idea of enclosing the tube containing the liquid oxygen in some container which aimed to prevent the inflow of heat from the room-temperature conditions of their laboratories. Placing a drying agent within the outside container absorbed water from the air and prevented frosting, leaving them able to see what was going on in the innermost tube. But none of the devices was outstandingly successful.

Once oxygen has been liquefied, the scientist finds that Nature works to help him keep it liquid, because the latent heat of vaporization is quite considerable—70 calories are required to vaporize one cubic centimetre of liquid oxygen. If heat can be prevented from getting at the liquid oxygen, then the scientist can keep litre quantities fairly freely in his laboratory for experiments on the behaviour of other forms of matter at these low temperatures, or for demonstrating dramatic effects to large audiences.

Dewar solved the problem sometime in 1892 and demonstrated his vacuum-insulating flask in a discourse on 20 January 1893. The principle is simple—a double-walled glass tube with a vacuum between the two walls. Since there is virtually no matter between the two walls, very little heat can be conducted across. In our domestic vacuum flasks we normally use the principle to keep the contents of the inner tube hot, because the heat from the tea, coffee or soup cannot be conducted away to the outside world across the vacuum—but the same principle means that heat from the outside world cannot be conducted to anything very cold within. By silvering the inside walls heat transfer can be further reduced by cutting out the effects of radiation. Dewar had apparently used vacuum insulation twenty years earlier to insulate a calorimeter in which he was carrying out experiments on heat, and he turned back to the idea when he wished to preserve cold. He demonstrated the effectiveness of his insulation in his usual dramatic fashion—having shown his audience the bluish liquid oxygen sitting quietly within the double glass walls of his vacuum flask, he broke the outer wall at the point where it had been sealed, allowing air to rush into the insulating space, and immediately the liquid oxygen started to bubble and boil furiously until soon it had all disappeared.

Needless to say, Dewar's claim to have invented the new device was challenged immediately. The Germans said that a man named Weinhold had used vacuum insulation first; the French coined the title 'vase d'Arsonval' after their own nominee for priority. But Dewar's claim has prevailed, and justly so, according to the evidence available to us; every low-temperature scientist nowadays refers casually to 'the Dewar' when he wishes to indicate the vessel containing liquid air or liquid nitrogen or even liquid helium.

By 1895 oxygen and nitrogen, the two principle constituents of our air, had been liquefied and potentially commercial processes for this liquefaction had been patented. Now a veritable race to liquefy hydrogen was on.

By this time physicists were not wholly in the dark about the conditions they would have to achieve to liquefy hydrogen, since the Dutch theoretician Johannes Diederich van der Waals of Leiden had contrived to explain the shapes drawn on Andrews's 'maps'. The classical theory of gases was based on the assumption that the molecules of gases were so small and so far apart that interactions between them could be ignored. What van der Waals did was to find a simple way of expressing the fact that the molecules do, in fact, interact and attract each other. Thus he was able to account properly for departures from Boyle's law which experimentalists like van Marum had observed in the liquefaction of ammonia. He realized that by studying the magnitude of any gas's departure from Boyle's law at temperatures well above the critical temperature it was possible to predict what the critical temperature of that gas would be (if it were ever possible to reach it). For hydrogen the value to be expected for its critical temperature would be about minus 240 degrees—and this was ultimately confirmed to be the correct value.

It soon became clear that there was a big gap between the lowest temperature that could be obtained by pumping off the vapour from a bath of liquid nitrogen and the target figure of minus 240 degrees. Liquid nitrogen could be got down to minus 210 degrees by pumping off the vapour, but at that stage the nitrogen froze solid and ceased to be an efficient coolant. So there was little hope of liquefying hydrogen by any variation on the cascade process. Dewar decided that the Joule-Thomson cooling device offered the best chance of dealing with hydrogen. But his first efforts were extremely frustrating, for on first passing hydrogen through the porous plug the gas started to heat up rather than cool down.

But van der Waals' theoretical work provided an explanation for this unexpected discovery. His work showed that there was another special point, the so-called inversion temperature, in the description of any gas; if a gas was put through the Joule-Thomson device starting above its inversion temperature it should get hotter—it would only cool if it was started below its inversion temperature. And van der Waals' work showed that this inversion temperature should be just six and three-quarters times greater than the critical temperature—which meant in the case of hydrogen a temperature of minus 50 degrees. Plainly it would be easy to cool hydrogen below this temperature with liquid air.

Dewar had got these preliminary difficulties out of the way by 1895, the year when Linde and Hampson patented their machines for liquefying air. It still took him until 10 May 1898 to liquefy hydrogen, by first cooling it with liquid oxygen and then passing it repeatedly through a Joule-Thomson porous plug until he had about twenty cubic centimetres of colourless liquid sitting quietly in a vacuum flask, boiling only slightly.

He formally announced his success at the Royal Society on 12 May. Strangely enough, he never formally described his apparatus and only gave some technical details about precooling the gas and the rates of flow. There is, however, no reason to doubt that he succeeded as he said, for he was able to repeat the process regularly and he later achieved further fame as the first man to solidify hydrogen.

Dewar's first experiments with liquid hydrogen ran into some interesting and prophetic troubles. He tried to measure the temperature of the boiling point of hydrogen, i.e. the temperature he had reached in the process of liquefaction, but both his attempts to use electrical thermometers—a thermo-couple and a platinum resistance thermometer—failed completely. Eventually he had to use a clumsy apparatus that was really a version of the early gas thermometer and from his results he deduced, rather than measured, the boiling point of hydrogen as about twenty degrees above absolute zero—very nearly correct.

Dewar's announcement of his liquefaction of hydrogen precipitated a controversy with Hampson over the credit for using the Joule-Thomson cooling system, but this subsided into angry mutterings after a furious exchange of letters in the scientific press. More serious in the long term was his row with Sir William Ramsay, one of the most distinguished British scientists of the day. This had begun late in 1895 at a Royal Society meeting when Dewar had described some of his early work towards the liquefaction of hydrogen. After he had finished by expressing high hopes about the eventual achievement of this objective, Ramsay stood up and said he had just heard from Cracow that Olszewski had already succeeded in the research. This left Dewar with little to do but carry on, and wait for the official news of his rival's success. No publication came from Olszewski, however. But when Dewar announced his successful result, Ramsay stood up again and proclaimed that Olszewski had earned the right to priority. This time Dewar challenged him to produce proof. At the next meeting of the Society Ramsay had to admit publicly that he had just received a letter in which Olszewski denied that he had ever managed to achieve a complete liquefaction of hydrogen. Ramsay had undoubtedly behaved strangely, but when he failed to put his statement into print, Dewar published the whole controversy and his own clear victory. Relations were then as bad as they could possibly be, and this was not just a petty squabble; it seriously interfered with the next major step towards absolute zero.

For, of all men to have quarrelled with, Ramsay was the world's greatest authority on helium. Helium had been discovered in the sun in 1869, revealed by its specific strong yellow line in the spectrum of the corona of the sun observed during an eclipse. Ramsay was the first to find helium on earth, twenty-five years after it had been located on the sun, and in his laboratory, within walking distance of the Royal Institution where Dewar worked, he had the greatest expertise and the best equipment for separating helium and the other rare gases from the atmosphere. Dewar thought at first that he had liquefied helium at the same time as he had liquefied hydrogen. He turned out to be wrong and he realized quickly that hydrogen was not the last great step towards absolute zero. Helium remained to be liquefied, the last gas to resist the scientists' attempts to prove Lavoisier correct. Plainly its boiling point must be even lower than that of hydrogen.

Ramsay needed Dewar's liquid hydrogen to help him in his quest to separate helium from neon. Dewar needed Ramsay's large supplies of helium to enable him to work towards liquefying the gas. But neither would even speak to the other, much less cooperate. And so the liquefaction of helium was first performed in Holland, at the University of Leiden, where Kammerlingh Onnes had built up the largest and most powerful cryogenics laboratory the world had ever seen.

The town and University of Leiden hold a peculiarly central position in the story of low-temperature science. Van Marum, the first man to liquefy ammonia gas, worked at the University. Van der Waals was a native of the town, though he was at the

University of Amsterdam when he published most of his significant work. The cryogenics laboratory built up by Heike Kammerlingh Onnes in the last years of the nineteenth century, after he had been appointed professor of physics at Leiden University at the age of twenty-nine, was to be the scene of many of the most important discoveries in low-temperature science, and was to dominate the new science for the first quarter of the twentieth century. But, more than that, this new laboratory was to set the pattern for the big research departments that we see today. Kammerlingh Onnes was the first of the 'big business' professors, the man who saw that science could no longer be an amateur affair, that major discoveries could no longer come from a single man working in his own one-room laboratory with devices held together by plasticine and elastic bands. He set the pattern of the professor as the man who runs the team, the collector of money, the winner of grants.

Kammerlingh Onnes did all this because he was forced to by the very nature of the science that interested him. In order to study matter at low temperatures, in order to liquefy gases that no one else could liquefy, he realized that he needed large quantities of liquid air and machines that would make these large quantities regularly and reliably. Later he would need large quantities of liquid hydrogen, with the machines that went with that. He would need large and powerful pumps to produce high-vacuum conditions for long periods of time. To run these pumps and liquefiers he would need a large team of mechanics and technicians. Such men were not readily available, so he would have to train them himself. Men like Sir James Dewar were highly skilled craftsmen themselves, but the Dutchman saw that this was not enough; he would need highly skilled glass-blowers, for instance, to manufacture experimental apparatus such as no one had needed before. He set up his own school of glass-blowing whose products were later to be found in virtually every laboratory of note.

Kammerlingh Onnes, in fact, set up first as an industrialist rather than a professor of physics: a man running a large technological establishment, rather than the stereotype of the devoted scientist. To do all this he needed money, and he needed a certain amount of political muscle. He had no hesitation in setting about building up a ground base providing both these commodities—

and many of his contemporaries were somewhat shocked by his 'advertising'. He selected a large empty space in the middle of Leiden for his laboratory—it was an area blasted clear by the accidental explosion of a gunpowder barge on the canal. He carefully gave half the area as a public park, but he soon ran into local political trouble when residents discovered that he would be using compressors and compressed gases such as hydrogen in his laboratory. Being somewhat sensitive to the possibilities of explosions, they demanded a halt to the building and a public enquiry. They got both, and Onnes's work was held up for two years while he mobilized scientific and political support to impress the enquiry.

It was ten years after his appointment as professor that Onnes was able to start building the large cascade type of plant for liquefying air and oxygen that was the foundation of his work. But because of his careful planning and meticulous design the plant started work promptly in 1894, without teething troubles, and was large enough and reliable enough to supply all the needs of the world's largest cryogenics laboratory for thirty years.

Despite being plagued all his life by poor health, Onnes still behaved like a nineteenth-century paternalistic employer in industry. Tales are still told about him in Leiden—how he would regularly visit any of his technicians or glass-blowers if they were ill, riding round in his carriage. But he was careful to note on these visits any of his men with new carpets or furniture, and he would point out that they were doing very well and had better not come to him for any increase in their wages in the near future. He made them work hard, including all day Saturday, but he worked the same hours himself. Everyone at the laboratory went on their annual holiday at the same time, and came back at the same time —and that included the professor himself.

On top of all this he was a first-rate scientist, a man of the most meticulous accuracy in all his experiments. At a time of revolution in physics, when much work was qualitative rather than quantitative, he based himself upon the motto, '*Door meten tot weten*'—through measurement to knowledge. And in the course of his scientific life he produced an enormous amount of first-class work including some of the most significant discoveries in the history of science, opening up the whole field of work close to the unattainable absolute of low temperature.

D

It was also a feature of Onnes's approach to science that his laboratory and his work were open to everyone. There was none of the jealously guarded mystery of the work of Dewar and Ramsay. Which was as well for science, since if anyone wanted to do work on low temperatures in the first quarter of this century they virtually had to go to Leiden to do it, because no one else could offer the facilities. It is, for instance, not widely known even among scientists that those pioneers of radioactivity, Becquerel and Pierre Curie, both went to Leiden to study whether radioactive emanations from the newly discovered naturally radioactive elements, uranium and radium, varied at low temperatures. To their great surprise they discovered there was no decrease in radioactivity at very low temperatures. This was one of the most important clues to the fact that radioactivity is a property of the atom itself and its structure, and not some concentration of extrinsic energy.

Perhaps one can get closest to this extraordinary man, Heike Kammerlingh Onnes, in the History of Science Museum in Leiden, which contains many of the devices and pieces of apparatus he used. There are also documents in his own handwriting, such as his work on the letter to van der Waals. This was a purely private letter, not for publication, but the two men were co-operating on the thermodynamics of gases; van der Waals's work on drawing out more accurate maps of gas behaviour was of vital importance to Onnes, for the maps predicted where the critical points of gases as yet unliquefied were likely to be found. Onnes's original draft of his letter covers two foolscap sheets, each a nightmare of crossings-out, corrections, insertions of later results, and rewritings. An enormous labour of care and meticulous thought, resulting in a final letter to his collaborator of four smaller octavo sheets of neat handwriting.

The Leiden team did not liquefy hydrogen till eight years after Dewar first succeeded in this. But when they did so it was with a machine which produced a steady flow of as much as four litres an hour. This almost commercial scale of production was needed to supply the large apparatus that Onnes now built for his attack on the last of the gases to resist liquefaction : helium. The apparatus is preserved in the older part of the building that is still the cryogenics laboratory of Leiden University. No piece of delicate glass-blowing this—a tangle of sturdy pipes and valves, though now disconnected from the pumps and supply tanks which it needed in its heyday. And it was only two years after the opening of the supply of liquid hydrogen that the machine achieved its great success.

The crucial day was 10 July 1908. Work started at 5.45 a.m. with the seventy-five litres of liquid air which had been prepared the previous day. There was no certainty of success when they started, for though Dewar had predicted that the critical point of helium would be somewhere between five and six degrees above absolute zero, a point they might reach, Olszewski had calculated that it would be only two degrees above zero, and Onnes's own earlier work supported Olszewski. However, the latest calculations made at Leiden, taking gases down to liquid hydrogen temperature, gave hope that Dewar's prediction might be more correct. Throughout the morning of 10 July they used the liquid air to liquefy hydrogen, and with this they started the afternoon by cooling their apparatus. Onnes's own paper—*Leiden Communication Number 108*—gives full credit to his famous chief technicians, Flim and Kesselring. It was Flim's job to get the hydrogen through all the pipes of the apparatus without entrapping any atmospheric air which would have frozen solid and either blocked the pipes or frosted up the glass of the central container, making observation of the behaviour of the helium impossible.

It was late in the afternoon, however, before they were able to introduce the helium into the central part of the system. They had little to guide them on what would actually happen when they entered the regions well below the boiling point of hydrogen—the final ten degrees above absolute zero. A helium gas thermometer had been placed in the central cryostat, but no one really knew how this would perform at these extreme temperatures. Otherwise they could only watch and hope they would see some clear change.

For an hour or so they fiddled with different adjustments of the valve through which the helium was expanded, and with variations in the pressure. Then the temperature in the central vessel appeared to start falling, according to the gas thermometer. But it did not go down steadily as in normal cooling, and soon it stopped going down at all. It was now 7.30 in the evening; the liquid hydrogen was almost all used up; they had been working for more than twelve hours non-stop. Apparently they had failed.

But the knowledge that something big was being tried had spread round the Leiden University grapevine, and various visitors had dropped in to see what was going on. At the moment of apparent defeat one of these kibitzers, Professor Schreinemakers, suggested that perhaps they had succeeded after all, but just could not see it. He proposed they should illuminate the central vessel more brightly. To do this they had to shine a light from below the central cryostat. And as soon as they did this, the light from below showed them clearly that there was liquid present. The reflection from the meniscus, the surface between the liquid and gas, was obvious. They had finally succeeded in liquefying the last gas.

But in their triumph they did not forget to observe like good scientists. It was noticed in particular that the meniscus of liquid helium had an unusual appearance—unlike other liquids it appeared to be weakest or thinnest exactly where it reached the edges of the glass tube that made up the central cryostat. This was the first hint, though they did not appreciate it at the time, of the very odd behaviour of materials at very low temperatures, and it was the first demonstration of the properties that make liquid helium one of the happy hunting grounds of the most advanced physics experiments in our own times, seventy years later.

Kammerlingh Onnes was never a man to stop driving. As soon as it was clear that he had liquefied helium, he threw his remaining resources immediately into an attempt to reach the final stage, solidification of the gas. He employed the last of his liquid hydrogen; he allowed the liquid helium to evaporate until there were only a few cubic centimetres left; and then he turned on his strongest pumps to draw off the helium vapour and reduce the temperature even further. He got no further result. (We know now that in this first great liquefaction of helium Onnes must have reached a temperature only one degree above absolute zero, but he never solidified helium during his life and he could not understand why.)

Finally, about 10 p.m. on that historic day Onnes called a halt. A very human note sounds in *Communication 108*:—'Not only had the apparatus been strained to the uttermost during this experiment and its preparation, but the utmost had also been demanded from my assistants.'

The long road to the total liquefaction of air had at last reached an end. That end was no blank wall—it turned out to be the start of several new roads—the road towards absolute zero, the road to the new technologies of the super cold and the road that was to be followed by the new industrial gas industry.

5 Cold Gas

The largest industrial use of low temperature and cryogenic engineering techniques is in the production and transport of gases. All the gases we use can be handled most conveniently, can be separated from each other, transported, and stored most easily, when they have been turned into liquids. The liquefaction of gases is usually achieved by making them very cold.

Every chemical element and most of the 'natural' chemical combinations that we know on earth can appear in three phases—solid, liquid and gas. The phase depends on the temperature and pressure, but if we assume the pressure remains the same, then any substance is solid at some low range of temperatures, changes into liquid at some higher temperature, and changes into gas at some temperature even higher still. Gold, iron, sand, all the things we think of as solid, can be turned into liquids, and gases as we raise their temperatures. Water is liquid at normal temperatures, but changes into solid as the temperature falls below freezing point, or changes into gas as the temperature rises above boiling point. Likewise substances which we know as gases can first of all be liquefied and then solidified as we take their temperatures steadily lower. (Helium is the only substance which will not turn solid at a low enough temperature; we have also to increase the pressure to get solid helium, but this has been achieved.)

The temperatures at which substances boil or freeze (keeping the pressure constant) are specific to each substance and depend on its basic structure, on the nature and content of its atoms. These temperatures are so specific that they can be used to identify each substance—if we have some mysterious lump of metal we can heat it until it starts melting and identify it from the temperature at which melting occurs. But 'melting point' also means 'freezing point'—a metal at a very high temperature is a liquid, and as it is cooled it solidifies at some temperature. This is its freezing point, and the temperature is, of course, the same as that at which the metal melts when it is heated. Metals similarly have a boiling point at which the liquid becomes gas.

Because different substances have different, and highly specific, boiling points, engineers can use the fact to separate substances which are mixed together. The mixture is heated and the substance with the lower boiling point (called the 'lighter' or 'more volatile' substance) turns into a gas which can be drawn off from the liquid mixture. If the original mixture is a complicated one a further increase in temperature will cause the ingredient with the second lowest boiling point to be driven out of the liquid so that it, too, can be separated. Such a process can obviously be extended to provide a way of separating the ingredients of the most complex mixtures. And it is, in fact, the principle upon which the giant oil refineries operate to provide so many of the basic ingredients of our modern technological civilization. Crude oil is split up into the petrol and benzene fractions to drive our cars; at a higher boiling point naphtha is separated to provide the raw material for many plastics. Then comes the heating oil which keeps many of our homes warm; then the diesel oil for lorries and heavy industrial engines; then the waxy paraffins; leaving a residue, which is almost solid, and which can be used as asphalt for road surfacings and roof coverings.

The idea of a boiling point, a temperature at which liquid turns into gas or vapour, is familiar to us all, and we regard it as the result of applying heat to the liquid. But that same 'point' can also be regarded as the 'liquefaction point' in a gas which is being cooled. Water in its vapour phase, which we call steam, becomes liquid again when it is cooled, and we call this process of liquefaction condensation. Therefore, just as we can separate the components of crude oil by heating the oil and making use of the differing boiling points of the various components, so we can separate the components of air by making the mixture even colder, and taking advantage of the differing 'boiling points' of the components to separate each one as it condenses and turns into a liquid.

The air around us, our atmosphere, has many components. Nitrogen is much the largest proportion of the air, 78%. Although nitrogen is an essential component of all living matter because of the part its atoms play in amino-acids, enzymes and all other proteins, it is comparatively inert in its gaseous form and does not easily combine with other atoms. (The difficulty of extracting nitrogen from the atmosphere and persuading it to combine with other chemicals to form biologically useful substances is the primary constraint on all plant growth and is therefore the major limitation on our ability to grow more food.) The second main component of air is oxygen, which makes up 21% of the atmosphere. Oxygen is highly reactive; its combination with other atoms, which we usually describe as burning or corrosion, is the chief mechanism by which man transfers energy from one place to another, either through using the heat of a flame or by breathing in oxygen and using it to 'burn' with food in order to supply his own energy of movement.

The remaining 1% of air is mostly made up of the 'rare' gases, argon, neon, helium, krypton and xenon, each of which has its own uses. But there is also a certain amount of carbon dioxide in the air—about 0·035%, and traces of hydrogen, nitrous oxide and ozone. Our atmosphere also contains large amounts of water vapour, which can be thought of as steam at low temperature, and this varies according to local weather conditions such as temperature, pressure and humidity. As much as 4% by volume of our atmosphere can consist of water vapour when conditions are very humid, but this does not affect the composition of the air, which is for practical purposes the same at all times and all places at average ground level.

The gases of the air all have different boiling points—which means that they all become liquid at different temperatures as the temperature of the whole is taken downwards by refrigeration. Xenon, the least important, becomes liquid at minus 108°C. Oxygen liquefies at minus 183°C and nitrogen at minus 196°C. Helium has the lowest boiling point of any known substance, minus 269°C, only four degrees above absolute zero.

In the industrial process of separating the different gases in air, it is not possible to achieve practical results simply by cooling to the boiling point of one gas and attempting to obtain a pure product from the liquid at that point. The process used in practice is more like liquefying the whole and then separating the different components as they turn back into gas when the temperature is raised again. This process is called distillation, but any single application of the technique results, at the appropriate temperatures, in a gas which is 'richer' in oxygen than the normal composition of air, and a liquid which is richer in nitrogen.

The standard industrial air separation plant consists, in principle, of two columns or tall towers. In the first, use is made of the fact that, at pressures about five times atmospheric, the boiling point of nitrogen is higher than the boiling point of oxygen at atmospheric pressure, and this range of temperature is easier to achieve than the minus 196°C of liquid nitrogen. This stripping column divides the air into two fractions, an oxygen-rich fraction and an almost pure nitrogen fraction. These two fractions are fed into a 'rectification' column, which contains many levels of metal trays pierced with large numbers of small holes. The whole is kept at a very low temperature and a stream of nitrogen-rich vapour flows upwards. The flows are balanced so that the vapour coming up through the holes in the trays stops the liquid from dripping down, and the liquid descends from one tray to the next through pipes. The vapour coming up into each tray is warmer than the liquid on the tray and causes the liquid to boil. There is therefore a small distillation at the level of each tray and so the vapour gets richer in nitrogen as it goes up while the liquid becomes richer in oxygen as it falls down.

To drive this process, it is necessary to keep the column warmer at the bottom than it is at the top. There is therefore a reboiler at the bottom to keep a flow of vapour going upwards, and there is a condenser at the top to remove heat by liquefying part of the rising vapour, and keep up the supply of downwards-falling liquid. Both liquid and vapour are drawn off from the top and bottom of the column, and by careful arrangement it can be secured that the liquid from the lower end (not from the very bottom of the column) is pure oxygen, while pure nitrogen is recovered from the top. The oxygen in particular can be recovered 99·7% pure.

This description is still over-simplified, for much of the skill in designing an economical and efficient air-separation plant lies in recirculating the cold products in such a way that they are used to cool the incoming flows, thus reducing the amount of power needed in the refrigeration plant. There are also the important extra 'columns' for the production of the rare gases.

Because their boiling points differ from those of oxygen and nitrogen, three of the rare gases become most highly concentrated at different points in the oxygen-rich flow of the main column. A small proportion of this flow is bled off at these points and this side-flow may be found to contain as much as 20% of argon, say, in otherwise almost pure oxygen. In an 'argon enrichment column' the oxygen may be steadily distilled off in a smaller version of the main rectification column until a product that may be 98% pure argon is drawn off. The remaining small percentage of oxygen can be removed by forcing it to react with hydrogen in the presence of a platinum catalyst. But further distillations may still be needed to remove the minute proportions of hydrogen and nitrogen before really pure argon emerges. Krypton and xenon are drawn off the main column at a different point, and this side-stream has to be sent to yet another column for 'krypton and xenon enrichment'. Again the basic process is the same, but in the final product the krypton has to be separated from the xenon by absorbing it on to charcoal at the temperature of liquid nitrogen.

Neon and helium have boiling points even lower than nitrogen, so these gases concentrate in the liquid nitrogen of the main column. Some of this liquid nitrogen is drawn off especially for helium and neon production, and the two rare gases are separated from this liquid by boiling away nitrogen at rather high pressures, where its boiling point becomes much lower. Then the helium and neon can be separated from each other by absorbing them on to different materials.

All these complicated operations are nowadays performed in vast industrial complexes. To take but one example, in 1975 the British Oxygen Company won the largest contract in its history—and the company is ninety years old—for supplying oxygen and nitrogen worth £50 million to steelmaking plants in the Teesside area in the north of England. The contract called for the daily supply over fifteen years of 1,200 metric tons of oxygen and 500 metric tons of nitrogen. The gases can at first be supplied from BOC's existing plant in Middlesbrough, the largest industrial gas plant in the UK with a daily output of more than 5,000 tons of gas. But to meet the contract in later years, the company will build an additional plant which will produce more than 2,700 tons of oxygen and nitrogen daily. This additional plant will cost around £15 million and will make the Middlesbrough gas-producing complex probably the largest in Europe. The oxygen and nitrogen will be sent by pipeline to blast furnaces and oxygen steel-making plants both south and north of the River Tees. Both gases are also supplied to the heavy chemical plants of ICI Limited in the same area; oxygen in large quantities is supplied to a factory making titanium pigments, and nitrogen is delivered to a chemical factory on Teesside owned by the international Monsanto company. The pipeline grid in the area supplies smaller quantities of gas to the shipbuilding and engineering industries lying nearby, and it is expected that the refineries and petrochemical plants planned for Teesside as the North Sea oil flows into the local terminals will be additional customers.

Very large air-separation plants such as this are often nowadays constructed primarily to serve new steel plants. The American-based company Air Products has, for instance, recently completed a plant with a daily capacity of 1,500 tons of oxygen, adjacent to the British Steel Corporation's plant at Llanwern in South Wales. Large plants of this sort are invariably called 'tonnage' plants in the industry.

The industrial separation of gases by liquid distillation began in England, France and Germany in 1902, after liquefaction on a large scale had been pioneered at the end of the nineteenth century by Hampson, Claude and Linde (see Chapter 4). Several of the great industrial gas companies of our day descend directly from these three men. The French company L'Air Liquide claims Georges Claude as its progenitor. In Germany the Linde company still dominates from its base in Munich, and the industrial gases section of the American giant, Union Carbide, is called its Linde division. In Britain Hampson's work started the British Oxygen Company on its way to a share in the world market.

There are, however, ways of producing oxygen other than by liquefaction of air and its subsequent distillation. In 1851 the Frenchman Boussingault discovered a reaction which involved heating barium oxide in air to 500°C when barium peroxide is formed by adding an oxygen molecule taken from the air. Further heating to 800°C releases oxygen and regenerates the barium oxide molecules. Unfortunately this chain soon breaks unless the air is cleaned of carbon dioxide, dust and organic detritus. The two Brin brothers discovered this and also improved the process further by introducing pressure changes. They started manufacturing oxygen in this way, but in 1886 the name of Brin's Oxygen Company was changed to the British Oxygen Company and in the first years of the present century their process, too, was taken over by the air liquefaction method.

Yet another way of producing oxygen is by applying electricity to water so as to dissociate the hydrogen and oxygen atoms. The process is called electrolysis. In principle, electrolysis is expensive in terms of power used, and it is not well adapted to a large-scale production, tonnage production, of gases. But electrolysis can be performed in small cells, and these cells can even be made as mobile units, so that oxygen can be manufactured exactly where it is needed. In the late 1930s an oxygen salesman in the American Middle West, Leonard Pool, realized that the transporting of heavy steel cylinders filled with compressed oxygen from a central plant to customers scattered over the vast distances of the USA was uneconomic. For many orders the cost of transport exceeded the cost of the gas. With the aid of a young engineering student at the University of Michigan, Frank Pavlis, he built a small oxygen generator that could produce 350 cubic feet of the gas an hour. They assembled the unit in the corner of a building owned by a Detroit transport company and set out to build a business based on the idea of providing oxygen generators at the sites where the gas was needed. Thus was born, in 1940, Air Products Inc. The war made the company into a big success. Companies like Union Carbide were reluctant to sell gas-making equipment to the government and the armed forces; they visualized moving centrally-made gas all over the world in cylinders. Leonard Pool, however, was well placed to sell gas generators that could go on to individual ships or be joined together on trailers to make mobile oxygen units for the army. By the end of the war Air Products was selling five million dollars worth of equipment a year, but had no civilian commercial business at all.

At this stage the revolution of oxygen-steelmaking came to the USA, pioneered by the Weirton Steel Company, a West Virginia division of the National Steel Corporation. Gas delivered in steel bottles could never meet the tonnage demand for oxygen of a plant that needed four hundred tons of gas every day. Air Products, with its tradition of providing oxygen at the place where it was needed, got the contract in 1947, and started the new trend in building oxygen plants adjacent to steelworks, using its own money to build the plant, and supplying oxygen by pipeline for revenue over the years. The contracts with Weirton, and soon afterwards with Ford Motor Company, gave Air Products a new foundation. The Korean War provided more contracts in the military field, and the arrival of the missile and space programmes, with their enormous demands for liquefied gases, ensured a vastly increased market for the company.

Perhaps because of its origins and comparatively short history, Air Products has diversified into other branches of industry more than most of its rivals. It has major interests in the chemicals and plastics industries and has even penetrated into chemical engineering plant manufacture. This has led to interests in the fuel and nuclear power industries.

The French giant L'Air Liquide believes firmly that its job is to sell gas. Although it will research into, and pioneer, new fields of development, it will only do so if the new field will generate increased sales of gas. Even in research the Frenchmen make no bows to Anglo-Saxon intellectual snobbery; there is no talk of 'pure' research for the sake of gaining knowledge in its huge low-temperature laboratories at Grenoble. Will it lead to more economical production of gas, easier transport of gas or sales of more gas? Those are the questions asked.

From its beginning in 1902, building on the work of Georges Claude, the company has steadily increased until now it probably holds the leading place in the world's industrial gas industry. Among its historical triumphs it includes the separation of neon in 1908, the production of argon in 1914, a synthesis of ammonia in 1918 and the separation of krypton in 1928. Right from the start it tried, and succeeded, to get work outside France, and the company claims to have advanced particularly in the USA in the 1970s. It now has 330 plants spread over the five continents of the world. It employs more than 25,000 people, and the 1973 sales were more than $3\frac{3}{4}$ billion francs. Like all its rivals, L'Air Liquide has been forced to go into the business of producing equipment that uses its gases, such as welding equipment, and similarly it has had to go into chemical engineering plant manufacture. It has interests in plastics, chemicals and electronics, but nearly 60% of its activity remains the sale of gases.

British consumption of oxygen in 1938 was 36,800 tons. With the ending of the Second World War it jumped to 126,000 tons in 1948. The million tons a year mark was reached in the early 1960s. In 1972 it was 1,840,000 tons and it is now around the two million ton mark. Not all of this has been provided by the British Oxygen Company, but the company has expanded in tune with the market till it now has sales of £300 million a year. Around the world it has naturally built up its strength in the former British Empire countries and it has not yet penetrated the US market to any extent. In the 1950s the company diversified into other industries more than its French counterpart, but this policy has been reversed in recent years. It has left BOC, however, as one of the largest providers of computer services in the UK through its 'bureau' operation. The company is also involved in a fish-farming venture, but the reason for this is that their system involves blowing oxygen through the fish tanks to get the fish to grow faster.

The original discoverer of oxygen, Joseph Priestley, tried breathing his new gas and recorded that his 'breast . . . felt peculiarly light and easy for some time afterwards'. He warned that it would have to be used carefully in case it caused the physiological processes of the body to 'live out too fast' but he thought it might still 'prove very useful as a medicine'. As individuals this is probably the use of oxygen that we are most likely to encounter—its medical use. The production of medical gas supplies has usually remained the province of the first company to operate in any country. It is not a commercially attractive market and the distribution problems, involving every hospital in the country and even the homes of individual patients, are enormous, while the safety and security checks are expensive though absolutely necessary with patients' lives at stake. Nitrous oxide, the original laughing gas, is still the most widely used anaesthetic. But medical gas supply is not the sort of market that an industrial gas-producer will choose to penetrate when it is expanding away from its home base. In some of the largest hospital complexes it is quite common nowadays, however, for a central supply of oxygen to be kept in liquid form in a large insulated tank from which the gas is boiled off and circulated by pipeline.

The basic trend that has caused the growth of these giant international companies has been the growth in the demand for oxygen—primarily for steel-making but also for welding, metal-cutting, chemical and plastics manufacture and a host of other uses. The growth in the demand for the by-products of oxygen production, the 'rare' gases, has been increasing in parallel. Argon can hardly be called a 'rare' gas any more—consumption in the UK alone has been of the order of 20,000 tons annually in the 1970s—and it is nowadays considered more accurate to refer to these gases as 'inert' gases. Argon, in fact, is used in those processes where neither oxygen nor nitrogen can be tolerated. The most important of these uses are to provide the gas filling for electric light bulbs and to provide a 'blanketing' atmosphere to protect molten metals from the air during many welding processes. It is also used to provide an inert atmosphere in the manufacture of highly reactive metals such as zirconium and titanium or in the finely-controlled processes of manufacturing transistors based on silicon or germanium.

Neon is the gas that provides the bright lights in the city—its own familiar red and orange range of colours, or the blues, greens and lavenders that can be obtained by mixing it with different proportions of argon and helium. Xenon and krypton are even more satisfactory fillings than argon for electric light bulbs, but because of their scarcity and comparatively high price they are only used in special applications. Xenon is a useful mild anaesthetic, but rarely used as such because of its price.

Neon, however, has one particularly interesting use, which is in the giant 'bubble chambers' used by nuclear physicists to detect the sub-nuclear particles which emerge from their enormous particle accelerators. (A large quantity of liquid neon and hydrogen is allowed to expand for a brief period of time until the gases are almost ready to boil, and then the particles from the accelerator—protons, neutrons, electrons and many others which go to make up atoms—are allowed to shoot through the chamber, marking their passage so that scientists can identify them by thin lines of bubbles as the liquid gases boil just where they pass.) Liquid neon is so rare and expensive to produce that the quantity required for one of these bubble chambers is worth millions of dollars, since there is only one kilogram of neon in 76 tons of air and its boiling point is minus 246°C.

The laboratories running the largest nuclear accelerators in the USA have therefore reached an agreement to share their rare and valuable liquid neon supplies which are the greater part of the American national stock of liquid neon. And in January 1976, 11,000 litres of liquid neon was transported from the Brookhaven laboratories on Long Island, New York, to the Fermilab at Batavia in Illinois, many hundreds of miles away. There had been more than a year and a half's work on the safety aspects of such transport before the big move began, because the liquid would later be needed in the bubble chambers of other laboratories as well as back at Brookhaven.

A group of Fermilab had made special modifications to a Dewar trailer—a gigantic, wheeled, vacuum flask—and a tractor designed by the Argonne Laboratory hauled the trailer. Two journeys had to be made, with a driving team of two men on the vehicle and an escort car with relief driver following. A steady speed of 45 m.p.h. was the objective and each trip was scheduled to last about 26 hours. But the move ran into some of the worst winter storms of the year. On the second journey the escort car was lost but the drivers of the main vehicle chose to push on over the icy roads and completed the last several hundred miles by themselves.

Although more special equipment and stricter safety precautions are necessary for carrying neon in liquid form, the advantages outweigh the extra effort. It would have required twenty or more standard vehicles to transfer the neon as a gas.

This is another reason why the liquefaction of air dominates the industrial production of pure gases. The transport of oxygen, nitrogen, hydrogen and other elements is vastly more economical when they are in liquid form, even though refrigeration or specially manufactured Dewar vessels are required to keep temperatures low. The standard steel 'gas-bottle' which we see in industrial welding or in a hospital contains oxygen compressed to 132 times atmospheric pressure. Such a cylinder weighs 140 lb when empty or 160 lb when full, and contains 240 cubic feet of gas when it is let out at ordinary pressure and temperature. The bottle therefore weighs eight times as much as the gas it carries. For hydrogen, because the gas is less dense, the transporting cylinder weighs 110 times as much as the gas it carries.

On the other hand, a ton of oxygen occupies 26,540 cubic feet at normal temperature and pressure, but occupies only 31 cubic feet when liquid. Putting the same figures in terms of transport containers, the vessel needed to carry liquid oxygen weighs only about half as much as the oxygen it carries—compared with eight times the weight for carrying oxygen as a gas even when it is highly compressed.

It is therefore common to use transport tankers carrying ten or eleven tons of liquid gas, and larger vessels holding thirty tons are coming into employment. A central liquid oxygen tank holding supplies for some large industrial undertaking may well hold 1,500 tons which is equivalent to 40 million cubic feet of gas. The traditional steel gas cylinders are often still the most convenient method of handling gas at the point of use, but pipeline systems, carrying either liquid or gas, are rapidly becoming more common, especially for distribution from a central store or liquefaction plant.

But as air-separation plants have increased so much in number and size around the world, there has arisen the problem of disposing of the liquid nitrogen which is produced as a necessary by-product in the rectification columns of the oxygen-producing plants. The drive by the industrial gas companies to find uses and markets for this by-product, instead of allowing it to escape to the atmosphere, has been the chief motive power behind the rapid spread of cryogenic techniques into many branches of the food processing and engineering industries.

Nitrogen has always had its uses, the principle being to combine with hydrogen to make ammonia, the basis of fertilizers and explosives. It has also been used as a 'blanket' of comparatively inert gas to cover molten metals or glass during manufacture to protect them from the reactive gases in the remainder of the atmosphere. But the discoveries that liquid nitrogen at minus 196°C preserved substances such as blood and cattle semen more effectively than ordinary refrigeration showed the gas companies a route to sales of the large quantity of liquid nitrogen they were producing. It is the pursuit of this line of sales-development that has led the industrial gas companies into the cryogenic food business, taking over a development that might have seemed to come more naturally from those companies already established in the frozen-food business.

The problems of 'selling' liquid nitrogen are threefold. First, new uses for the material, or for the extreme cold it provides, must be created or invented. Secondly, the potential users must be converted to cryogenics from whatever may be their present practices. Thirdly, the liquid nitrogen must be made available on a wide scale. In the first of these problems, as we have seen, scientific research has shown, quite independently of the gas companies in many cases, that greater cold is advisable in itself. The discoveries that blood and semen are better stored at lower temperatures is the prime example. The second problem has to be overcome by the normal techniques of salesmanship.

The creation of nationwide systems of liquid nitrogen delivery is one of the main preoccupations of the gas companies at the moment. Major improvements in the design and manufacture of insulated road and rail tankers have been involved. Anyone, anywhere, can light a fire and provide a heat treatment method for dealing with some problem or process. To persuade people to adopt what may be the more efficient techniques of cryogenics to solve their problem or to carry out their process depends on being able to offer a reliable supply of 'cold'. We are all accustomed to the idea of energy, or potential heat, being supplied to every building by gas-pipe, or oil-hose, or electric cable. We cannot supply cold so easily. It must at present be brought as liquid gas in insulated containers by some form of wheeled transport. Some of the gas companies are building up fleets of small trucks, each carrying an insulated 'pot' of liquid nitrogen to supply the small user. The driver of the van is also a trained cryogenic technician, expert in the uses and applications of liquid nitrogen techniques. (More of the new uses of liquid nitrogen will be described in Chapter 8.)

Liquid helium and liquid hydrogen are not produced by the air-separation process. Both these elements are present in the atmosphere only in very minute proportions, although hydrogen itself is one of the commonest elements. Furthermore, hydrogen liquefies (or boils) at minus 253°C and helium at minus 269°C, only four degrees above absolute zero. The temperatures of air-liquefaction plants never get as low as these figures. Yet liquid helium and liquid hydrogen are both important in the high technology world, and therefore the techniques of low-temperature engineering and science have to be used to get the two elements, normally gaseous, into liquid form.

Hydrogen is normally produced as a pure gas by the electrolysis of water (which we have seen is the process of using electric power to break water up into its constituents of hydrogen and oxygen) or as a by-product of the production of chlorine by the electrolysis of brine. Hydrogen gas is quite widely used in the production of synthetic ammonia or in the plastics industries—British consumption was half-a-million tons annually at the beginning of the 1970s.

Liquid hydrogen, however, has only one important use at the

moment, and that is as a rocket fuel. Because hydrogen is the lightest of all the atoms it has, theoretically, the highest specific thrust as a fuel—which means that any single atom of hydrogen expelled backwards from a rocket motor provides a greater forward thrust to the rocket for its weight than an atom of any other element. This property of hydrogen was recognized by the pioneers of rocketry such as Robert H. Goddard, who suggested as far back as 1923 in a Smithsonian Institution publication that hydrogen and oxygen were the ideal rocket fuels. It was not until 27 November 1963 that the USA launched the first successful rocket stage propelled by liquid hydrogen and oxygen—this was the NASA Centaur rocket, powered by two hydrogen-oxygen burning rocket motors. It went up propelled by an Atlas booster rocket, before performing successfully on its own. Subsequently, liquid hydrogen and liquid oxygen were selected as the fuels for the upper stages of the mighty Saturn rocket vehicles in the US moonflights.

We know nothing about Russian production of liquid hydrogen, but it can safely be assumed that only the USA and Russia are large-scale manufacturers of liquid hydrogen. At the peak of NASA's assault on the moon, at the end of the 1960s, US production of liquid hydrogen was exceeding 2,000 tons a year, and there were five plants operating, the three largest of which were in California, with the Sacramento plant capable of turning out 50 tons a day.

By a rather strange concatenation of circumstances, the pioneer of large-scale liquid hydrogen production was the US National Bureau of Standards. The NBS has a long history of research into low-temperature physics and the special technologies required, since its normal line of work is to establish the facts about the properties of materials at all temperatures and pressures. For this reason, it was asked by the Atomic Energy Commission in 1950 to try to design and develop a hydrogen-liquefying plant many times larger than anything existing in the world at that time. The work could not be done at NBS's historic site in Washington, DC so a special research laboratory was built at Boulder, Colorado, and there, in 1952, a hydrogen liquefier with an output of 300 litres per hour was started up. But huge areas of ignorance were immediately opened up, too, for very little was known about the behaviour of materials on an engineering scale at the temperature of liquid hydrogen, minus 253°C. Work had to be started on the storage, handling and transport of liquid hydrogen, as well as discovering such things as how ball-bearings would behave in an environment only twenty degrees above absolute zero.

Towards the latter part of the 1950s, however, the interest of the Atomic Energy Commission in liquid hydrogen steadily decreased, and today the nuclear industry has no practical use for the material, though in the future the liquefaction of hydrogen may be the best way to provide supplies of deuterium and tritium (hydrogen$_2$ and hydrogen$_3$) for a large-scale programme of electric-power stations based upon controlled nuclear fusion. But just as the interest of the nuclear industry waned, the rapidly expanding space industry became convinced of the value of liquid hydrogen as a rocket fuel. So the programme of the Boulder Laboratory continued and it was renamed the NBS Cryogenic Engineering Laboratory. It also established a national data centre for all cryogenic technical information.

Some experts believe that liquid hydrogen will one day play a vital role as the chief energy storage medium and a widely used fuel in a world energy system no longer reliant on virtually exhausted supplies of fossil fuels. This theory will be more fully explained in the final chapter.

At present liquid hydrogen has one other important use, apart from driving rockets : in the production of liquid helium. Because hydrogen's low boiling point is nearest to helium's, cooling with liquid hydrogen can be an important step in getting down to the extremely low temperatures needed to persuade helium to turn into a liquid.

All commercial supplies of helium in the western world come from natural gas wells in the USA and Canada. In a very few of the natural gas fields helium forms as much as 3% or even more, of the outflow, and this is much the most abundant source of helium, for there is little of the gas in the atmosphere and there are no helium compounds in the chemistry of the natural world.

Liquid helium has only two major uses. Because of the very low temperature needed to make helium liquefy—minus 269°C—the liquid can be used in the production of extremely low temperatures in other materials. Liquid helium is used in all laboratories investigating phenomena that occur as absolute zero is approached. The practical application of such extremely low temperatures is only just beginning to appear, but is so important, especially in the production of superconductivity, that a special chapter will be devoted to it.

The only truly industrial use of liquid helium at present is as a method of providing large quantities of helium gas for use in deep-sea diving. Just as with other gases, it is easier and more economical to transport and handle helium in its liquid form than as a compressed gas. Even the extremely powerful refrigeration equipment and extensive insulation required to provide supplies of liquid helium are worth the outlay in view of the necessity of helium to the offshore oil industry of the world. In order to transport liquid helium it must be contained in a Dewar vessel of its own, surrounded by liquid nitrogen in yet another Dewar vessel, yet this is the only reasonable way to transport the large quantities of helium needed by divers working many hundreds of feet below the surface of the sea. The diver needs an atmosphere of more than 90% helium if he is to breathe at great depths—the rest of his atmosphere contains the oxygen he needs for life, but he cannot have nitrogen in his breathing mixture for at high pressure nitrogen will dissolve into his blood and affect his brain so as to intoxicate him. As the western world struggles to find oil and produce it from ever deeper waters in the North Sea or off the eastern coast of North America, the work of divers is essential to our economies. So cryogenics has become an essential technology.

The same reasons that are forcing the western nations to produce oil from under the sea are causing us to reorganize our entire system of gas supplies. Natural gas, broadly speaking, comes from the same places that oil comes from; but while we can ship oil in tankers from the Middle East or North Africa across the Atlantic to the USA or across the China Sea from Borneo to Japan, we cannot in economic or practical terms ship natural gas across the oceans, or pipe it either. The answer to the problem lies in liquefying natural gas by making it cold, and then shipping the liquid just as we do oil.

Natural gas consists mostly of methane, a combination of carbon and hydrogen atoms, which has been known for centuries in the coalmining industries as the dangerously explosive gas 'firedamp'. It also contains varying, but small, proportions of other hydrocarbons such as butane and propane. Virtually all the world's oilfields contain natural gas, for oil, too, consists of hydrocarbons of various types. But there are some deposits of natural gas which provide no oil.

In technological terms the liquefaction of natural gas stems directly from Carl von Linde's development of the air-liquefaction process—the same principles underlie the process and the same sort of machinery is used. And the first liquefaction of natural gas was performed with the object of separating gases by fractional distillation, just as in the case of air liquefaction. However, the object in the first liquefaction of natural gas was to separate and produce helium from the helium-rich natural gas wells of the American Middle West. LNG (liquefied natural gas) was first produced in 1924 by the US Bureau of Mines as a by-product of the drive to obtain helium in large quantities for lighter-than-air filling of gas-bags of airships. (The hydrogen used to lift the airships of the First World War and the years immediately following had proved to be dangerously inflammable.) By reducing the temperature of the natural gas to minus 162°C, all its hydrocarbon components liquefied (the boiling point of methane is minus 161·6°C and the lesser components have higher boiling points), leaving the required helium as a virtually pure gas.

Interest in lighter-than-air craft diminished rapidly through the 1920s and 1930s as the aeroplane proved more efficient and economical, and no more was heard of LNG for nearly fifteen years. But during those years the use of natural gas as a public utility fuel was growing steadily in the USA and the networks of pipelines and facilities spread out from the natural gas wells.

The major economic problem of all systems of public fuel and energy supply is the problem of meeting peak demand. For just a very few days of each year—the coldest days—the demand for energy reaches a peak, far higher than at any other time. Providing the heavy capital equipment—power-stations and transformers, gas production facilities, pumps and pipelines—to meet this peak demand in full is enormously expensive, because the equipment remains unused for the rest of the year. For the gas industry, local storage is the obvious answer, so that gas made or produced in the summer can be kept near to the place where it will be needed in the winter peaks. Obviously storage tanks are an extra expense, and the smaller they can be made the more economical they are likely to be. If natural gas is compressed to 70 times the atmospheric pressure, the volume is reduced by a factor of 70. The same quantity of natural gas, when liquefied by cooling, is reduced in volume 580 times—so comparatively small tanks allow large quantities of gas to be stored to meet peak demands.

The idea of using liquefied natural gas for the 'peak-shaving' operation was discussed in the USA throughout the 1930s but it was not until 1940 that the first liquefaction plant was built. In the years of the Second World War several more such facilities were built, but in 1944 there was a disastrous accident and fire at a Cleveland plant and all further progress was stopped for twenty years. It was not till the 1960s that further liquefaction plants for peak-shaving were built, but there are now some fifty in the USA alone.

During this period, when liquefaction plants were regarded as unsafe, there was a most important development. To overcome the safety problem, the possibility was considered of putting the liquefaction plant on moored barges in the middle of Lake Charles, Louisiana. Then the LNG was to be shipped by barge up the Mississippi to Chicago. Nothing came of this scheme, which was first proposed in 1952, but some of its ideas were incorporated into the first major transportation of LNG in 1959, when the specially modified crude-oil tanker *Methane Pioneer* carried several cargoes from Lake Charles to Canvey Island, the British Gas terminal in the Thames Estuary.

The idea of supplying huge quantities of gas to a system by transporting LNG was thus born—supplying base-load gas in this manner is very different from peak-shaving, and requires different, and much larger, facilities. The *Methane Pioneer* went on to start a regular service of LNG supplies from a huge liquefaction plant at Arzew in Algeria, using natural gas from the Algerian oilfields, to the Canvey Island terminal, from which it was fed into the British Gas national pipeline grid. Although this system demonstrated to the world an entirely new method of supplying energy from distant fields to a heavily populated, industrialized, gas-consuming community, it was not to prosper itself. For shortly after the pioneering venture had begun, large supplies of natural gas were discovered under the sea-bed of the southern North Sea, and it is these fields and not the Algerian oilfields, that now supply the British public with most of its gas.

The rest of the world took note, however. By the middle of the 1970s, with the world oil crisis upon us, LNG was being shipped out of Alaska, Algeria, Libya, Borneo and Abu Dhabi, to the USA, to France and Italy and to Japan. And at least 25 other major schemes were being planned or actually built throughout the world. Nigeria, Russian Siberia, Ecuador, Trinidad, Iran, Pakistan, Bangladesh, Indonesia, Sarawak and Qatar were all planning to become exporters of LNG destined mainly for the USA but with significant quantities also going to Japan and Europe.

The whole of this development is well summed up in *Liquefied Natural Gas* by W. L. Lom (1974). He has been associated with Exxon's natural gas operations for many years and he writes: 'Until quite recently gases produced in remote regions of the world were either considered valueless, and the wells from which they had been produced were plugged and abandoned, or if natural gas production was associated with that of oil, the "associated" gas was either reinjected into the oilproducing formations, or it was flared. Clearly the burning off of large volumes of gas, however unavoidable, was frequently considered unacceptable by local administrations and alternative uses had to be found for the associated gas. Pressure was exerted on the producing companies to convert gas into petrochemicals and later, when this became technically feasible, to initiate gas liquefaction and LNG shipment projects. In other words, initially such projects were adopted largely to appease local objections to the flaring of large volumes of surplus gas.'

But the coming of the oil crisis and the ever increasing demand for energy have changed that situation. Lom continues: 'While the first LNG projects were initiated as much to provide clean fuels to cities polluted by smoke and other forms of emission as to mitigate shortages of locally produced gas, we note now that LNG is being imported and will be even more so in the near future mainly to supplement local gas sources which are no longer adequate. Not only is gas demand increasing very rapidly in all the industrialized countries but, as a result of this large demand, local gas sources are being exhausted at an unexpectedly rapid rate; LNG is therefore being called upon to meet energy shortages in general and local gas shortages in particular.'

The harsh facts behind these statements are that the North

American gas reserves (as calculated in 1971) are only sufficient to provide 13 years' supply if the demand continues at the 1975 level. Western Europe has gas reserves for 29 years' consumption, while the Soviet Union has 47 years' supply in reserve. Russian and Iranian gas is already being piped to Western and Eastern Europe through systems stretching for hundreds of miles in length. But no pipelines can cross the Atlantic Ocean or the seas that surround Japan. The political developments of recent years have served only to emphasize these points.

The facts, then, make LNG shipment projects desirable from the point of view of both the producing countries and the potential consumers. But LNG transport has the additional advantage of being much more flexible than pipeline schemes, because the ships can sail to different terminals than those originally intended if either supply or demand dries up at any particular place—and this seems more likely to occur for political than technical reasons in the near future. Indeed, it is suggested by Lom, among the criteria desirable for setting up a large LNG shipment scheme that 'While relations between producing and receiving countries need not be cordial, they should at least be correct and the probability of a revolutionary change of regime in either which could lead to the cancellation of the project should be low'.

The capital investment in an LNG shipment scheme is enormous. First, the gas must be piped from inland or underwater fields to the large liquefaction plant which is ideally on the coast. A harbour capable of taking tankers of 100,000 or 200,000 tons is necessary, and the loading facilities must be elaborate for safety reasons since the liquid methane can be dangerous. At the receiving terminal there must be equally elaborate unloading facilities and there must also be insulated storage tanks. Finally, there has to be a re-gasification plant feeding the methane into the local gas-supply system.

Many new technologies have had to be developed before LNG shipping projects have become practicable. Most notable among these has been the design of the LNG carrier ships themselves. At the temperature of liquid methane—minus 161°C—ordinary 'mild' steel from which most ships' hulls are constructed becomes brittle and is liable to sudden catastrophic fractures. (The embrittlement of steel at low temperatures was learned the hard way through a number of disastrous failures of Liberty ships during the Second World War.) Even the smallest leak of LNG from the tanks of the ship on to the plates or ribs of the main hull could therefore be disastrous. Metals which will stand the low temperatures of LNG without suffering major physical changes or becoming brittle are aluminium and some of its alloys, stainless steel, and steel with at least 9% nickel in it. The tanks containing LNG as well as all the piping and structures which might come into contact with LNG must therefore be made of these materials. Furthermore, all the LNG-containing structures must be insulated to keep the LNG cold and liquid. This can, however, be used in a 'trade-off' manner in the construction of LNG carriers, for insulation would have to be provided whatever material was used for the construction of the tanks and hull of the ship.

Two alternative methods of constructing LNG carriers have evolved. In one system, special LNG tanks are fitted into a conventionally constructed hull, with the tanks virtually free-standing within the body of the ship. Each tank is insulated separately and there is free space around the tanks for inspection against leakage. In later developments of this method, the insulation has been used to key tanks into their positions in the ship. The advantages of this system are that each tank can expand and contract by itself, without straining the whole structure, and the whole is presumably safer in case of collision or impact accident. The first LNG carrier, *Methane Pioneer*, simply had separate LNG tanks built into its original structure as an oil tanker. All subsequent LNG carriers beginning with *Methane Princess* and *Methane Progress*, have been specially designed and built for the job. Aluminium was used for tanking in all earlier ships, but later carriers have used waffle or honeycomb stainless steel tanks.

The disadvantage of free-standing tanks is that they waste a lot of hull space. The construction of LNG carriers in which the tanks are an integral part of the ship began in Sweden, working to French designs, providing ships for the Alaska–Japan run. In these ships the insulation—often perlite in plywood boxes—is laid against the inside of the main hull. The tanks are constructed of double layers, with insulation between, of thin membranes of special stainless steels such as Invar, which contains as much as 36% nickel and which hardly expands or contracts at all. These tanks are connected to the main structure of the ship and follow the shape of the external hull. Despite the greater economy in space, and the greater ease of construction in shipyards, there are doubts about the safety of integral-construction ships, and both the ships on the Alaska–Japan run in the mid-1970s suffered accidents which caused LNG leaks.

The development of LNG carrier ships is still going on. By 1972, 13 had been built, the largest of which could carry more than 71,000 cubic metres. At the same time some 24 more of these special ships were under construction, with the largest aiming to carry 125,000 cubic metres of LNG. Orders placed amounted to more than 60, with the Onassis line planning monsters capable of carrying 165,000 cubic metres of liquid methane. But for many of these orders the shippers were not expecting delivery till the last years of this decade, and the methods of construction had not yet been revealed. There is even technical speculation that it might be feasible, and most economical, to build large LNG carriers of reinforced and pre-stressed concrete, which, in theory, should be less affected by low temperatures, and should provide better insulation, than any metals.

Although the scientific principles involved in the liquefaction of natural gas are exactly the same as those for the liquefaction of air and its component gases, the technology used in large gas liquefaction plants tends to be rather different. Production plants for LNG tend to be of the cascade type, depending less on compression of the gas and more on reducing its temperature by a series of steps using other liquefied gases as refrigerants. These refrigerants are most commonly minor components of natural gas, such as butane and ethane, which liquefy at higher temperatures than methane, the main component of natural gas. A cascade liquefaction plant has a complicated flow pattern, with compressors and expanders for each 'step' down the cascade removing the heat from the incoming natural gas first by heat-exchange with liquid butane, then by heat-exchange with liquid ethane, until a final compression-expansion cycle liquefies the methane. Alternative cascade systems use liquid propane and liquid ethylene, and there are even systems which use refrigerants of the freon type, such as are used in domestic refrigerators.

The storage of LNG has also involved the development of new technologies, at least where very large quantities have to be kept to supply base-load gas for an entire system. The 'frozen hole' concept is the most dramatic of these new ideas. In this system the whole area of ground earmarked for storage is frozen to a depth of many metres by putting refrigeration pipes into the earth. When the whole volume is frozen solid, a large cylindrical excavation is made in the middle. This is capped by a concrete 'lid' and the LNG is fed into the excavated space. The low temperature of the LNG will keep the earth around it frozen solid and thus a 'tank' has been created from a mere hole in the ground.

In practice the 'frozen hole' concept has not been very satisfactory. There have been unexplained leakages of heat into the frozen earth; the supports for the concrete lid have been pushed out of place by 'delayed frost heave' in the frozen earth; and the cyclical rise and fall of the water level underground as the tides rise and fall is believed to have been the cause of cracks appearing in the solid-frozen earth, for most LNG storage systems are naturally built near the port of entry. Double-skinned metal tanks, with insulation between the skins, have been very successful in storing LNG, but just as in the carrier ships, the tanks have to be made of aluminium or expensive steels that will not become brittle at low temperatures. Once again, reinforced concrete, with the reinforcing rods stressed under tension, is proving a satisfactory material for tanks, although there still has to be a thin metal container inside the concrete to ensure that the tank is gastight.

There are still unsolved problems in the storage of LNG in large quantities—mainly in the shape of a phenomenon called 'roll-over'. In all storage tanks some of the LNG is being drawn off for use by the public and the tank is regularly refilled or topped up before it has been completely emptied of its previous contents. This can result, if the fresh supplies are pumped in near the bottom of the tank, in a layer of colder, denser methane forming a stratum above a layer of slightly warmer, less dense LNG. For unknown reasons the two do not always mix well, but the higher temperature of the newly added liquid causes much vapour to form. Eventually the colder, denser liquid will fall to the bottom of the tank and the warmer 'rolls over' on to the top.

This leads to the release of even more vapour and there is an enormous rise in pressure. A very serious accident nearly occurred in a tank containing 50,000 cubic metres of LNG and all the local accident and emergency services had to be put on standby, with local roads closed and all shipping warned to stand off.

Only after eighteen hours did it finally become clear that the emergency safety valves and flare-off system had coped with the crisis and the tank had stood up to the strain. But not enough is yet known about roll-over and the problem is being studied even in university laboratories.

The other problem with LNG is that of dealing with any spillages that may occur, especially while at sea. Methane forms an explosive mixture with air when the concentration of methane is anything between 5 and 15%. If exposed to air, LNG naturally takes some measurable time to warm up, evaporate into its gaseous form and mix with air. The danger after an LNG spill is therefore downwind of the spill itself—and the bigger the spill the further away will be the dangerous area where the mixture of air and methane is explosive. It may be 800 feet from a big spill if the wind is blowing at 5 m.p.h. But in addition, it has been observed that LNG, when spilt on to water, causes a series of explosions at or near the surface. There seems to be some interaction between the water, which is comparatively very warm, and the liquid gas at minus 161 °C. It is by no means certain what process is taking place, but it is thought likely that it is similar to the exploding of droplets of water falling on to a red-hot metal, where the vapour envelope round a still liquid centre of the droplet collapses with explosive force. Large-scale tests at sea have been carried out, but detailed results have not been made public. Further research is almost certainly required to solve the problem of LNG spillage at sea.

The principle remains, however, that the shipping of liquefied natural gas from distant oilfields to the main centres of population and industry is one of the most rapidly expanding features of the energy economy. And the liquefaction of fuel gas, along with the liquefaction of air, are the most important current applications of cryogenic techniques on an industrial scale.

6 Freezing Life

Of all the branches of low-temperature science, cryobiology was the first to start and has been the slowest to develop. Throughout the history of written observations of nature there have been strange tales of birds, animals, fish and insects found totally frozen, yet restored to full normal life on being thawed out from their icy prisons. Since the first Greek physicians of classical times there has been interest in the temperature of man's body, so obviously raised in a fever, yet quite clearly capable of being damaged by extreme cold as much as by extreme heat. Yet where fire consumed, it could be seen that cold and ice tended to preserve.

We have seen that the first scientists—the 'natural philosophers' of the seventeenth century such as Francis Bacon—carried out observations in the field. Probably the first recorded 'experiment' in cryobiology was performed by Henry Power in 1663. He took a jar of vinegar infested with tiny 'vinegar eels' and cooled it in a mixture of ice and salt until the vinegar froze and the little animals were, in his words 'incrystalled'. An hour or two later he thawed the vinegar and the creatures within it reappeared and 'danced and frisked about as lively as ever'.

From this, Power concluded that cold could not kill, as heat could. This was an old argument among 'natural philosophers'. Robert Boyle took the opposite view. He described two different ways in which the body could be killed by cold; the usual way was for the whole body to be slowly invaded by cold, with the victim becoming steadily drowsier and more comatose and finally dying apparently in a sleep; however, armoured knights on horseback in conditions of extreme cold were gripped round the waist by a band of cold, which caused them such torments and digestive troubles that they died of pain and exhaustion. Boyle observed that fish and frogs in cold water that was allowed to freeze survived quite well when the water was thawed again, but he believed he had seen small amounts of still liquid water surrounding one of the frogs. He also showed that fish frozen in ice for as long as three days did not recover when thawed out but were irrevocably dead. Boyle also saw that cold, as well as killing, could preserve—he himself preserved apples, eggs, meat, plant material and even mammalian eyes. He exposed a live rabbit to a bitter frost all night and found it survived; a dead rabbit exposed similarly to frost was frozen solid throughout its internal organs. He could do little more than publish his observations in 1683; he could extract no pattern or theory from them, and to this day his fame rests on his much more successful work on the effects of temperature change on gases which culminated in Boyle's law, the foundation of modern thermodynamics.

The winter of 1709 was particularly bitter in Europe, killing many animals and plants. Yet in the following spring the insects hatched out of their eggs at the usual time, exactly the same time as after a mild winter. A Dutchman, Boerhaave, who had been trying to measure the temperature of the severe frosts all winter and had managed to record as low as 14 degrees below zero, noted the phenomenon and wondered at it—thus opening what was to prove, nearly 250 years later, the most important aspect of cryobiology, the effect of cold on reproduction.

Reaumur, the rival of Fahrenheit in proposing temperature scales, was the scientist who first applied thermometry to cryobiological experiments, examining the reaction of caterpillars to freezing. He found that a certain species (we do not know which species since he did not name it) survived freezing down to the mark of minus 17 on his scale, which is minus 20°C. Caterpillars of another species were killed by this temperature, but would survive freezing down to minus 9° Reaumur. Then by cutting the creatures open he showed that, although on the outside they seemed frozen hard, he could still find some liquid inside their bodies if they had been cooled to a temperature from which they could recover. Creatures that were frozen right through did not survive, he found. Thus Reaumur seems to have been the first to discover that some insects survive cold conditions by what the cryobiologists now call super-cooling—they retain some liquid within their bodies at temperatures below the normal freezing point of their body liquids. This is one of the two main methods of cold-survival in insects and in many hibernating smaller mammals. (There are other groups of insects, it has since been discovered, that can survive total freezing.) Reaumur went on to show that the blood of different species of caterpillars froze at various different temperatures, and he likened these different bloods to different strengths of brandies, which also freeze at different temperatures. He showed, also, that the blood of birds and mammals froze at much higher temperatures than the blood of insects.

But Reaumur was primarily interested in temperatures as such, and so the claim to be the first cryobiologist probably goes to the Italian, Spallanzani, who published his main work describing his experiments in 1787. He systematically observed under the microscope the behaviour of the tiny animalcules which had been discovered in rain and gutter-water by Anthony von Leuwenhoek a century earlier. He watched them swim away from the steadily advancing ice-front as he cooled the water in which they lived. He saw that they were still quite happy in pockets of liquid at minus 6 degrees (he probably used the Reaumur scale). He saw that they swam more slowly at minus 8 degrees, and that most of them died at minus 9 degrees, even though there was still plenty of liquid water. And he found two species that survived these temperatures, though we do not know now which species these were. Spallanzani went on to show that some creatures, such as the tiny worms called rotifers, could be frozen to temperatures as low as minus 24°C and still survive and be thawed back to active life.

Spallanzani's interest was in the great controversy, which raged on well into the nineteenth century, about the nature of life and death. This included the problem of whether a creature could survive if all its life processes were brought to a halt by drying or freezing. Spallanzani believed his rotifers were examples of 'resurrection'. In pursuing his argument he put the eggs of butterflies, silkworms and several other types of insect into glass tubes, which he kept at minus 24 degrees (probably equivalent to minus 30°C) for many hours. Yet when the next spring came the eggs hatched out normally although the insects developed from the eggs were killed by temperatures of only minus 7 or 8 degrees. He showed that the dead insects were frozen right through but that the eggs contained some fluid inside even at the lowest temperature he could reach, and he believed that the eggs contained an oily material preventing them from freezing completely and thus preserving the embryo within. Thus he seems to have been the first to make the suggestion of some anti-freeze material which would protect cells from very low temperatures.

Finally Spallanzani seems also to have been the first to freeze sperm. He showed that frog sperm frozen for half-an-hour and then thawed out retained the power to initiate egg-development, though it lost this power if kept in the frozen state for several hours.

Throughout the nineteenth century and on into the first half of the twentieth century, further observations on the effects of low temperatures on living matter, and strange stories about the preservation of life under freezing conditions, continued to be reported. Some of these reports were very reliable. The Arctic explorer Ross found caterpillars surviving after being exposed four times to temperatures varying between minus 11 and minus 47°C. Franklin, back from his polar voyages in 1822, reported a carp leaping vigorously after it had been thawed from a block of ice in which it had been frozen for thirty-six hours. Other stories of humans being frozen accidentally until they appeared dead and then reviving when put in a coffin or even recovered from under a heap of Swiss ice are more closely linked to the dearly-loved human tradition of Rip Van Winkle and the Sleeping Beauty.

As time went on more truly scientific reports accumulated—many from naturalists interested in the phenomenon of hibernation, others from experts in particular fields, such as the life of yeasts, bacteria and other micro-organisms, who systematically tested their subject matter at a wide range of temperatures. The difficulty of this work lies in showing whether the creature is still truly viable after spending some period at a very low temperature, but nevertheless a considerable stock of knowledge grew up about the survival or destruction of micro-organisms at a wide variety of low temperatures. Furthermore, as the ability to produce much lower temperatures grew with the development of refrigeration technology, some scientists started to experiment with much deeper freezing. Pictet showed in 1893 not only that the eggs of the silkworm survived a temperature of minus 40°C, but that this treatment killed bacteria which had been attached to the eggs. He also cooled frog eggs to minus 60°C and thawed them successfully so that they developed into tadpoles.

But all this information and observation was without pattern or underlying theory of what happened to whole animals, tissues or individual cells, to unify it into some comprehensible model of the behaviour of living matter under the influence of extreme cold. Nor was there any scientist who might be called a cryobiologist; the observations came almost entirely from men whose main concern was some other subject and for whom the effects of cold were but one aspect of their other preoccupation. The late-eighteenth-century interest in the argument over life and death had gone, largely as a result of Pasteur's work showing that only life gave rise to further life.

The present concept of cryobiology can be said to have started with the publication of a book in 1940 which to some extent referred back to that ancient controversy, at least in its title—*Life and Death at Low Temperatures.* The authors were B. J. Luyet, a professor at St Louis University, Missouri, and his assistant, P. M. Gehenio. It may not exactly have been a case of the prophet being disregarded in his own country, but Luyet's work—essentially, the formation of some sort of pattern out of the vast, inhomogeneous mass of recorded observations on the effects of low temperatures and freezing on living things—proved seminal first in England, then on the European continent. And even now, though there are distinguished cryobiologists in the USA and in Canada, the greatest interest in the discipline remains east of the Atlantic.

The significance of Luyet's book was twofold—theoretical and practical. On the conceptual side, it showed scientists a new meaning in the old argument about life and death, namely that if one could ensure a complete stoppage of the metabolic processes of living cells or organisms without causing death, a state of suspended animation would be achieved which would allow for virtually unlimited storage of the cells or organisms. On the practical side, the book brought to wider notice work by Luyet himself, and others, in the very rapid freezing of cells, by which the water was 'vitrified', literally turned into a glassy substance, where the crystals of ice were extremely small. This, it seemed, prevented the growth of large ice-crystals inside or outside the living cells and protected the cells from the physical damage of actual disruption by ice-crystals.

The idea of rapid freezing to produce vitrification was followed up in the USA by a team of agricultural scientists under C. S. Shaffner who actually froze fowl and frog sperm to minus

79°C (the temperature of solid carbon dioxide), and showed that the sperm were still able to move (motile) after thawing. They apparently also retained some fertilizing power although no live chicks were produced. Shaffner had added the sugar laevulose to the sperm before freezing, with the object of partially drying the sperm so that there would be less water within them to turn into dangerous ice-crystals.

A French researcher, Jean Rostand, observed in 1946 that the addition of glycerol to frog sperm enabled many of the cells to recover from exposure to low temperatures of around minus 6°C, but this work was not followed up nor widely published, and its significance was not understood till much later.

The real beginning of modern interest in cryobiology came in 1949 as a result of a strange accident in the labelling of bottles at the Medical Research Council's laboratories in Hampstead and Mill Hill in North London. These laboratories were undergoing a post-war reorganization, including the rehabilitation of the main building which had been turned into a naval establishment manned (if that is the word) by the Women's Royal Naval Service. One of the new teams was formed by the endocrinologist Alan Parkes. Since pre-war endocrinology at the time of his education had been almost entirely concerned with the hormonal processes in reproduction, he had become interested in reproductive biology in all its aspects. He recruited Audrey U. Smith, a young woman with both medical and scientific qualifications, who had entered science despite some family disapprobation, and who had ended the war in the large team developing and applying penicillin. The third member of the team was Chris Polge, an agricultural scientist, who was given, among his other duties, charge of the experimental animal-breeding on the farm at Mill Hill.

Alan Parkes was a particularly brilliant man (he later became a professor at Cambridge University), notable for his wide vision and quick and exciting imagination. He believed strongly that truly worthwhile science should have some practical objective in view and he had the vision to see that successful long-term storage of sperm would revolutionize animal-breeding. 'Reproduction in mammals normally involves contemporaneous and contiguous action on the part of the two sexes. The advent of Artificial Insemination abolished this requirement in principle . . . but using semen long-stored in the frozen state has enormously extended the scope of the technique. It may be said in fact that we have abolished time and space in cattle breeding,' he wrote a few years after the great discovery of 1949.

It was Luyet's book that fired the imagination of Parkes and his team and they set about following the lead of Luyet and Shaffner in rapid freezing and vitrification after drying animal sperm by the addition of laevulose and other sugars.

They had no success at all. 'In our experience revival of fowl spermatozoa after vitrification is negligible when the technique of addition of laevulose is followed,' they reported. They had had considerable difficulty in getting supplies of pure laevulose under the immediately post-war austerity conditions, and when they did get some, they made up a large number of bottles containing the solution with the correct proportion of laevulose. These bottles were put into cold storage when their attemps to reproduce Shaffner's results failed. They then waited for further inspiration.

Towards the end of 1948 Chris Polge was worried about the main laboratory's demand for supplies of fresh experimental animals all the year round. The ability to store sperm in freezing conditions and thaw it out when needed would greatly help him to keep up a regular supply of animals, so he returned to the problem of attempting to freeze fowl sperm. He ordered the bottles of laevulose solution to be sent up to Mill Hill from the Hampstead laboratory cold store. Once again he had almost total failure when he thawed out the sperm after it had been frozen at minus 79°C. But from one sample he had what even the official scientific paper describes as 'dramatic results'. Almost all the fowl sperm became motile when thawed. This sample had been frozen in the solution from one particular bottle. He repeated the experiment using the solution from this one bottle. Again the sperm virtually all survived twenty minutes at minus 79°C. And, what was more, these sperm seemed to retain some of their fertilizing power as well.

At this time, so soon after the discovery of penicillin, the team's first thought was that a mould, observed in the bottle of solution, had acted on the laevulose to produce some substance which gave protection to the sperm against freezing damage. But tests on the solution soon showed that it contained no laevulose at all—in fact, it contained no sugar whatsoever.

By this time there was very little of the solution left in the 'magic' bottle—barely ten to fifteen millitres remained, about three teaspoonsful. This was handed over with some trepidation to one of the chemists, Dr D. Elliott, for analysis. One of the first tests Dr Elliott tried was to submit a tiny portion of the solution to a flame. Immediately he identified the familiar smell of burning glycerol. He soon identified protein and water as well, and immediately the scientists knew what was in the bottle. It is common in many biological laboratories and is called Mayer's albumen. It consists of 10% glycerol and 1% albumen (the protein) in water, and is used for sticking thin sections of material on to microscope slides. The team had been using it regularly in their histological studies of the sperm they had been trying to freeze. To this day no one knows how that bottle got mixed up with the stocks of laevulose solution—Chris Polge suggests the likeliest explanation is that its label had come off while being stored.

They very soon showed that albumen had no protective effect when cells were frozen and so it must be the glycerol that enabled living cells to survive at temperatures down to minus 79°C. The work was pubished in *Nature* on 15 October 1949, and the standard works on cryobiology acknowledge this as the starting point of the modern science.

The discovery that glycerol would protect cells at very low temperatures is one of the very few scientific discoveries that can be truly called a break-through. It opened up enormous new possibilities and virtually created a new scientific discipline. It also posed for the new science of cryobiology a problem which has bedevilled it ever since—the struggle between developing new applications and carrying out the basic scientific investigation to find out what is actually going on.

At first, however, the work of Parkes, Audrey Smith and Chris Polge was entirely devoted to developing practical techniques from their discovery. And it was far from a walkover. The fowl sperm which seemed so lively when thawed out seemed incapable of producing a live chicken. They eventually achieved only six live births out of 1,000 eggs fertilized by semen that had been frozen and thawed with a wide variety of different concentrations of glycerol and similar substances in an attempt to find the best way of preserving fertilizing power as well as life. It was only two years later that the answer to the problem was found by

Audrey Smith. Slowly all the possible causes of loss of fertilizing power were eliminated, and eventually she was able to show that the fowl sperm, although they were perfectly undamaged by immersion in strong solutions of glycerol at room temperature before they were frozen, and though they needed strong solutions of glycerol to protect them when they were at minus 79°C, were damaged by being exposed to these strong solutions of glycerol while they were thawing out. Chris Polge worked out the necessary process for removing glycerol by dialysis as the sperm was thawing out and before it was used for insemination. By 1951 they could produce whole clutches of live chicks using sperm that had been frozen down to minus 79 degrees.

Then came the question of how long the sperm could be stored when frozen. Immediately they received another blow, when they found that the sperm deteriorated quite rapidly and lost both its fertilizing power and its mobility after a few weeks' storage. They took the obvious step of storing at even lower temperatures—the temperature of liquid nitrogen, minus 196 degrees. They had no difficuty showing that sperm stored at this temperature could produce perfect chicks. Furthermore there was much less rapid deterioration of the stored sperm than at minus 79 degrees—again a situation in which the rule 'the colder the better' applies. But still there was some deterioration although it was possible to produce some chicks with sperm stored for more than a year. The long-term implications of this were plain to Alan Parkes: 'It could not be accepted that biochemical processes were going on at minus 190 degrees, and we had to accept the idea that damage due to physical causes might occur to cells during long-term preservation even at very low temperatures, and that the attainment of laboratory immortality would involve the suppression of physical as well as of biochemical changes.'

In the later stages of the work on fowl sperm, Audrey Smith's main interest was really elsewhere. Indeed, her vital discovery concerning the main problem, the damage caused to fowl sperm thawing out into a strong glycerol solution, arose from some even more important work she had done in the meantime. This was her discovery of how to freeze and store human red blood cells.

Immediately after it had been shown that glycerol protected fowl semen during freezing, and while the long struggle to find the right way of preserving a high proportion of fertilizing power was beginning, the team naturally thought of examining the effects of glycerol on the sperm of other species. Audrey Smith started with the familiar laboratory animal, the rabbit. But there was no good effect at all. Even in the very first paper this was reported: 'Results of a different order were obtained with rabbit spermatozoa. With this species no spermatozoa survive after bulk vitrification of semen.'

Trying to find the reason for the failure, Audrey Smith discovered that the rabbit sperm would not stand up to being placed in strong solutions of glycerol at any time. Under the microscope she saw that the delicate heads of the sperm, the acrosomes, became crinkled when they were exposed to glycerol and this immediately reminded her of earlier studies of other cells which had developed this crinkled appearance when placed in solutions which the biologist calls hypertonic, i.e. which have a greater osmotic pressure than the fluid inside the cells. (When there is a different concentration of salt on opposite sides of a porous membrane such as the cell wall, a chemical force known as osmotic pressure equalizes the salt concentration. When the salt concentration is higher outside the cell, water is drawn out and the cell shrinks; if the concentration is higher inside, water goes in and the cell expands.) Certainly it appeared as if the acrosomes of the rabbit spermatozoa had been dehydrated by the presence of glycerol.

To find out whether this was the case, Audrey Smith decided to use red blood cells (erythrocytes) from human and animal blood because these cells are exceedingly sensitive to salt concentrations and would therefore act as excellent osmometers to measure the situation in glycerol solutions. There came another surprise when the red cells were put into glycerol solutions. Even in strong solutions they shrank only slightly for a few minutes and then returned to their normal shapes and sizes. The obvious conclusion was that the glycerol had penetrated the membranes of the red blood cells and the concentrations inside and outside the cells had become the same. So perhaps glycerol protected fowl sperm during freezing because it could penetrate the spermatozoa of birds but not that of mammals such as the rabbit.

One question was solved immediately. Red cells protected by glycerol could be frozen and taken down to minus 79 degrees, stored for anything from five minutes to three months, brought back to normal temperatures and found to be intact and apparently vital.

The other question, as to whether glycerol has to enter a cell to protect it while frozen, or whether it is sufficient to have glycerol in the surrounding medium, is still not completely solved today, for although we know that glycerol can penetrate many cells, it probably does not enter all types of cell and the exact amount of penetration necessary to ensure survival is not certain. These questions, were, however, immediately pursued by two new recruits to the team, Henry Sloviter, a doctor from Philadelphia, and Jimmy Lovelock, who worked on the biophysics problems that cropped up all the time, the problems of getting machines and special equipment for performing duties such as the removal of glycerol or freezing and thawing at precise rates.

Sloviter took over the problem of finding whether frozen and stored red blood cells could be safely transfused into animals. Just as with the fowl sperm, it was found that the glycerol had to be removed from the red cells before they could be used again biologically. This was because, when placed in a saline solution or in blood plasma (the liquid part of blood), water was sucked into the red cells faster than glycerol could be expelled, so that the cells swelled and burst. But, again by dialysis, the glycerol could be removed at thawing and the cells could safely be transfused into a rabbit. Now the whole scale of experimentation had to be enlarged because the implications of blood storage were so vast. Sloviter and all the work went to the Medical Research Council's Blood Transfusion Unit, and Sloviter, of course, drew the attention of his American colleagues to his important discoveries. He quickly showed, by labelling his red blood cells with radioactive chemicals, that red cells which had been frozen and stored could be successfully thawed and, after the protective glycerol had been removed, could be put into blood plasma or saline solution and transfused safely into the blood stream of living animals. The frozen cells survived the transfusion as well as, or better than, fresh red blood cells and lasted quite as long as the standard hundred days' life of a red blood cell in normal circulation.

The chief problem to be solved before setting up banks of frozen blood cells was to reduce the number of cells killed during freezing and washing. Audrey Smith took no part in this work. After her basic discovery she was moved on to other things. The first experiments on freezing red blood cells were performed while Alan Parkes was on holiday. He was not pleased with what he found on his return. Red blood cells were dull and uninteresting to him—they do not divide, they have no active nucleus, they are not part of the reproductive system. He told Dr Smith firmly to start looking at the possibility of freezing the other important type of reproductive cell, the female egg cell, the oocyte.

The story of cryobiology consists very largely of the remarkable applications of low temperatures to living materials. The basic discoveries have nearly all been made by the members of Parkes's small team, while the developments have required large numbers of other scientists and, eventually, the help of whole industries. Telling the story of cryobiology thus becomes difficult, for one must look at each of the important branches of application of the science, and then return again and again to the main trunk line of basic discoveries, which have been made by a surprisingly small number of people over a very short time.

The first of the important applications of cryobiology came in the development of frozen red blood cell storage for transfusions. (Blood consists of three parts—the red blood cells, the various types of white blood cells, and the liquid plasma, which also carries a large number of important body chemicals in solution.) The basis for successful blood transfusion was laid by Karl Landsteiner in the first year of the present century when he discovered the ABO blood group system, and established that donor and recipient must have compatible blood for a safe and successful transfusion to take place. The first six successful transfusions were reported in 1906, but these were direct from donor to patient.

The demands of the First World War battle-casualties gave an enormous impetus to the development of transfusion techniques. By finding methods of sterilizing equipment such as tubes and bottles, doctors were able to collect blood and transfuse it in separate operations. Then it was discovered that sodium citrate prevented blood from clotting, and so it was possible to take from a donor and store the blood in bottles for as long as twenty-four hours. Nevertheless blood transfusion was still a dramatic and heroic treatment, reserved for extreme cases of injury or shock or the most severe blood diseases, right up to the years of the Second World War.

And again the demands of war provided the next major impetus to the techniques of blood transfusion. At the start of the Second World War there were a few blood banks keeping chilled blood for as long as a week at 5°C in Russia, Britain and the USA. But it was still preferable to use the blood within seventy-two hours of collection, and it was common in civilian practice to summon the donor to the bedside of the patient. The whole of our modern system of large-scale blood-banking and public campaigns for blood donors, and even most of the standardized equipment used in transfusing, were developed during the War. The discovery of acid-citrate-dextrose as a more efficient anti-coagulant and preservative then enabled blood to be kept for as long as three weeks.

But the 'twenty-one day tyranny' ruled the huge modern organizations for blood supply for nearly twenty years. The medical services remained shackled by their inability to preserve blood for longer than three weeks, until the discovery of how to freeze red blood cells could be turned into a practical nationwide system of blood banking towards the beginning of the 1970s.

Audrey Smith had shown in 1950 that red blood cells could be frozen and thawed. For the next five years Lovelock and Sloviter, and the many others who joined them in the work, struggled to find the best concentrations of glycerol to protect the cells, the best medium in which to put the glycerol, the best temperature at which to store the cells, and above all the best way of getting rid of the glycerol as the cells were being thawed. In 1955 it seemed that these tedious studies could perhaps be abandoned, for Harold T. Meryman and his co-workers at the Naval Medical Research Institute at Bethesda in Maryland showed they could successfully preserve red blood cells without any glycerol at all. They let droplets of blood, from a very fine spray, fall directly into a pool of liquid nitrogen at minus 196°C. The blood turned into a granular powder which slowly fell to the bottom of the liquid nitrogen, and when these granules were recovered and thawed they were found to contain intact red blood cells.

At this point industry, in the shape of the Cryogenic Division of Union Carbide, started to become interested. Many different techniques for immersing blood in liquid nitrogen were tried, of which the most successful was spreading a very thin film of blood onto pieces of stainless steel gauze and dipping the whole into liquid nitrogen.

These techniques have proved perfectly satisfactory for the freezing and storage of small quantities of red blood cells, and they are still used where only small samples are needed and where absolute sterility is not required, as in the storage of blood that will only be needed for serological laboratory tests.

But when it came to freezing pint quantities of blood, such as are needed for any major blood-banking scheme, the technique of direct immersion in liquid nitrogen ran into serious difficulties. Special flat metal blood-containers were tried, and then corrugated canisters were used in attempts to get very large quantities of heat out of the blood and into the liquid nitrogen as quickly as possible. The liquid nitrogen technique was essentially the same as the very rapid freezing and vitrification proposed by Luyet, and it proved impossible to get speedy enough cooling for large volumes of blood. The experimenters tried adding small quantities of 'cryoprotectants' such as glycerol or dimethylsulphoxide (DMSO, which in the intervening years had been found to work often as well as glycerol in protecting cells from freezing damage). Eventually cryoprotectants had to be used in concentrations as high as those in Audrey Smith's original experiments; the scientists were back with the old problem of getting rid of the glycerol when the blood was thawed out, and so the actual rate of cooling became immaterial.

The problem was not solved satisfactorily for pint quantities of blood till 1964, and even then the process was complicated and expensive. It involved putting the thawed but still glycerolized blood into a sterilized bowl revolving at 4,000 r.p.m. and steadily adding as much as four litres of solution containing steadily less and less glycerol. After ninety minutes of this treatment the red blood cells were still intact and most of the glycerol had been removed. It could then be shown that 90% of the frozen and stored red blood cells were active and useful when transfused into a patient.

The technique, with variations, is still in use, but it is expensive. It is estimated to cost £18 per wash, even with the latest equipment offered by the manufacturers which provide facilities such as continuous flow methods. Much of the cost covers the spinning apparatus (called a centrifuge) and the disposable bowls that are required if sterility is to be maintained. Other techniques which do not involve centrifugation have also been developed, using chemical reactions to separate the red blood cells gently from the glycerol-containing medium without allowing them to suffer osmotic damage, and then washing them clean. In these techniques the blood is in disposable plastic bags, and the system has been developed so that one technician can handle five blood-bags simultaneously and clean them all in twenty minutes. But this is still expensive in labour costs and in disposable materials.

The chief problem now is to achieve better economics and more efficient systems for freezing, storing, thawing and cleaning the red blood cells. The rapidly expanding medical supply industry is deeply involved in this development, with Union Carbide still in a leading position both scientifically and commercially. Whatever the detailed system used at any centre, deep-frozen blood is stored in insulated tanks containing liquid nitrogen, and so the industrial gas industry has a big interest in the blood-bank system.

In Britain, however, there is no widespread storage of blood in liquid nitrogen. Only the British Army has a large bank of deep-frozen blood, otherwise, in a geographically small, highly-populated island with a well-developed system of voluntary blood donation, it has not been felt necessary to install the expensive capital equipment and cryogenic facilities, though many doctors regret that we have not moved into this field and that they remain subject to the 'twenty-one day tyranny'. But in the wider spaces of the USA, deep-frozen blood banks have become common and the idea has been well developed in Europe.

One of the world's most advanced centres of deep-frozen blood storage is the Central Laboratory in the Netherlands Red Cross Blood Transfusion Service on the outskirts of Amsterdam. This is not only the central blood bank for Holland, storing most of its material in liquid nitrogen and administering a series of regional banks; it is also a leading centre of research into the problems of freezing and recovering red blood cells, and they have developed cleaning techniques which are among the most advanced in the world. Round this basic function have grown a number of other cryobiological research projects, including the preservation of other types of blood cells and work towards the preservation of whole organs.

The most advanced development of all, however, is the European Blood Bank which is part of this laboratory. The European

Blood Bank keeps only very rare blood types—the sort that may only be found in one individual among a million people. It services the whole of Western Europe, so that on the rare occasions when a patient needing a blood transfusion is found to have one of the unusual blood types, it can be supplied immediately to the hospital, which may be hundreds of miles away from the blood bank, across two or three national frontiers. The only alternative is to trace one of the few individuals known to have similar blood, and it is, of course, quite possible that there may be no known living person with an exact blood match.

The blood cells of these rare types are kept in paper straws, very similar to drinking straws. They are easy to handle and can quickly be taken from the central store and placed in a small Dewar flask for transport, probably by air, to the patient who needs the blood. Each straw is colour-coded according to the type of cells it contains, and they are stored upright in a grid of hexagonal-shaped stainless steel containers immersed in liquid nitrogen. A group of bulbous circular metal containers, each about the size of a large domestic washing-machine, constitutes the European Blood Bank. If you open the lid of one of the containers the mass of gaily-coloured paper straws can clearly be seen through what appears to be a thin cloud of smoke rising from the colourless liquid nitrogen. (The 'smoke' is actually water vapour from the atmosphere, condensing as it meets the intense cold of the gas above the liquid nitrogen.)

The problems of freezing, storing and thawing red blood cells, so that large-scale blood transfusion systems can be built on long-term-storage blood banks, have now been satisfactorily solved after more than twenty years work, although there is still work to be done on making the processes more economic. The freezing of other blood cells, particularly various types of white cells, is now one of the frontiers of cryobiology, and will be described later in this chapter.

The original break-through by Alan Parkes's team had been in showing that fowl sperm could be frozen, stored and thawed successfully back to life. But there is no great need for artificial insemination in poultry farming, still less in egg-production for human consumption. The need for artificial insemination and all the big problems of animal-breeding are found among the larger farm animals—cows, sheep and pigs—which have basically seasonal breeding patterns and long gestation periods, so that all experiments are time-consuming.

Naturally, after their dramatic success with fowl sperm, Audrey Smith and Chris Polge examined the effects of freezing and glycerol on other types of sperm. They had no success with rabbits, although we have seen that this work led to the discovery of red blood cell freezing. They also tried freezing bull sperm and had no success. But because the commercial importance of the storage of bull sperm would be so great they persevered with this particular line of experiment.

In the first days of this work the team was still influenced by Luyet's work and the concept of rapid freezing and vitrification to protect the cells from damage by ice crystals. After their first failures they realized it had already been discovered that bull semen was particularly susceptible to damage by cooling—indeed, it had been shown by earlier workers that it was all killed if the temperature was reduced to 0°C. There seemed then little hope that it would survive rapid cooling to temperatures even lower. They therefore started a long programme of protecting bull sperm with glycerol and cooling it very slowly. Many different rates of cooling were used, often with different rates above and below 0°C.

In nearly all cases, slow cooling followed by very rapid thawing enabled the bull semen to survive the journey down to minus 79°C. The largest proportion of sperm survived a regime in which cooling was done at the rate of one degree per minute down to minus 15 and then at five degrees per minute down to minus 79.

The theoretical importance of this discovery to cryobiology was as great as its later importance to cattle-breeding. The work showed that rapid cooling was by no means necessary to preserve cells at low temperatures as long as a cryoprotectant such as glycerol was used. Instead, the actual rates of cooling and thawing were shown to be one of the important factors in assuring the preservation of cells at low temperatures. This remains one of the most important fields of cryobiological experiment to this day, and in many laboratories the study of the optimum rates of cooling continues to be one of the keys to finding the way to freeze types of cell which have so far merely been killed by low temperatures.

As soon as the Mill Hill team had shown that they could preserve bull sperm at minus 79 degrees, they wished to see whether these sperm still retained the power to fertilize. The experiments were carried out for them by Dr D. L. Stewart, who inseminated five cows with semen that had been stored at minus 79 degrees for 48 hours using the Smith and Polge method. One of the cows became pregnant and duly delivered a live bull calf. This feat was achieved in 1951, and the control conditions to ensure that the animals had not been accidentally, but naturally, inseminated were strict, though strange. Each of the cows was housed separately at all times because they were being used for other experiments in animal physiology and veterinary medicine. In fact the cow that produced the first offspring from deep-frozen sperm was suffering from a fistula of its digestive organs.

At this stage Chris Polge, by qualification and by choice an agricultural scientist, left the team at the Medical Research Council Laboratories and went to the Cattle Breeding Centre at Cambridge to pursue the new line of work full-time. The larger-scale experiments he then started got off to a bad start. The appropriate control machinery to ensure the best rate of cooling was not at first available and the semen had to be cooled much more slowly, with bad results. Then another stroke of luck occurred. Some samples of semen were unintentionally left overnight, at a temperature of only 5°C, in contact with a solution of citrate buffer containing 10% glycerol. When they were rewarmed to 37°C, the sperm were seen to be exceptionally active. Some other specimens which had suffered the same accidental treatment were then frozen down to minus 79 degrees, and when thawed rapidly they, too, were exceptionally active.

This regime was then applied deliberately to fresh samples of semen, which were stored at minus 79 degrees for periods varying from two hours to eight days. Thirty-eight cows were inseminated with this material when it had been thawed. Thirty became pregnant, and in due course twenty-seven normal calves were born. The first, named Frosty, has achieved some permanent fame by having his picture included in the classical literature of cryobiology when the formal scientific report on their success was published in 1952. The name Frosty was bestowed on another calf twenty years later when yet another application of low-temperature preservation of living matter was pioneered by the same Cambridge workers.

A bank of semen frozen according to this new regime was immediately set up, as well as a system for automatically controlling the cooling rate to the speeds which had been found most satisfactory in the Mill Hill work. Polge and his colleagues were able to show that a year's storage did not reduce the fertility of the bull sperm by more than a small margin, and in due course they successfully impregnated 66.7% of their experimental cows with semen stored for $4\frac{1}{2}$ years.

There have, naturally, been many variations on Chris Polge's original techniques and materials, and many have given marginal improvements in the results. But within three years of the publication of his 1952 results, bull semen was being frozen, stored, and used for insemination in Australia, Belgium, Brazil, Canada, Chile, Denmark, France, Germany, Holland, India, Israel, Italy, Japan, Norway, Pakistan, South Africa, Sweden, Switzerland and the USA. Frozen semen had been exported from Britain to South Africa and New Zealand, while sperm from Canadian Holstein bulls had been flown, frozen, into Britain.

The Americans had always been rather ahead of the rest of the world in the application of artificial insemination to their cattle herds. This is probably due to the close relationship between the agricultural scientists at the American 'land colleges' and the farming communities among whom they live. But one man in particular was responsible for arousing American enthusiasm for the storage of frozen sperm—he was Rockefeller Prentiss of the American Breeders' Service, a commercial organization set up in the earliest days of artificial insemination just before the Second World War. He visited the research team at Mill Hill when the first experiments on slow cooling of bull semen were showing signs of success. Chris Polge remembers him saying, 'That's for me,' and he departed for America full of enthusiasm. Again Union Carbide was the pioneer of large-scale industrialization and the concept of frozen-semen storage was soon applied widely throughout the US. Naturally enough Union Carbide, as an industrial gas producer, soon developed the storage of bull semen in liquid nitrogen, rather than at minus 79°C in solid carbon dioxide. But European cattle breeders feel that the speed with which the Americans moved into large-scale operations has had certain disadvantages. The Americans are now committed, by their large capital expenditure, to the technique of storing the frozen material in glass ampoules of a fixed size. In the middle 1960s Britain and the other Europeans switched from ampoules to the use of paper straws to contain the semen. The straws are lighter, less fragile and easier to label and colour-code than ampoules. Furthermore, the tendency has been steadily to reduce the size of the straws, and the standard 'dose' is now only one quarter of a cubic centimetre, which has been proved more economical in that it freezes more easily than a larger quantity. But whatever size of dose is used, it should contain about twenty million sperm, for this has been clearly proved to be the optimum amount for ensuring fertility.

Three-quarters of all Britain's dairy cattle are now produced by artificial insemination with frozen sperm. The proportion of AI cattle is rather smaller in the country's beef herds, since much natural mating takes place among beef cattle ranging the mountains or hill country. The total number of artificial inseminations ranges between 2¼ and 2½ million a year. The success rate, as measured by the number of cows not needing a second insemination, runs at about 79%. It is believed that the conception rate is much higher, probably more than 90%, but the natural fertility rate of the animals is much lower than this, and it is therefore estimated that quite a proportion of pregnancies are lost for a wide variety of different natural reasons. Frozen stored semen has shown a higher rate of success than 'fresh' semen as originally used in the early days of AI, probably because the semen is far from fresh after it has been taken from the donor bull to all the cows requiring the service. Storage at minus 196°C, liquid nitrogen temperature, has also been proved more efficient than storage at minus 79 degrees.

Some smaller countries such as Denmark and Holland have an even higher proportion of AI cattle than Britain, but just as there are various differing techniques around the world for collecting, diluting, cooling, storing and thawing the semen, so there are varying needs and enthusiasms for the technique. In New Zealand, for instance, not all the AI practice is carried out with frozen sperm.

The freezing of sperm enables a good bull, with excellent or desirable genetic traits, to service far more cows over a far wider area than was ever possible before. He can even be servicing cows many years after his natural life has ended. It enables a truly scientific cattle-breeding system to be put into operation, so that the entire national stock of animals is improved. In Britain, for instance, the Milk Marketing Board is able to store the semen from a young bull while measuring the milk output of many of his daughters before deciding whether his genetic material will improve the national output and then releasing his semen on a large scale if the measurements prove satisfactory. Other factors, such as resistance to disease or rapid growth in beef breeds, can likewise be controlled by wide-scale breeding policies.

The ability to freeze and store sperm gives the cattle-breeders the chance to change the standard of a community's cattle stock at a speed which could never be achieved by ordinary AI and still less by any system of natural mating. Anyone who travelled casually around the British countryside in the 1940s would have seen that the cattle of the English landscape were, broadly speaking, brown, as in most paintings by British landscape artists. A similar journey through rural England today, only thirty years later, presents a different colour-scheme—the cows are black and white. The old dual-purpose beef-and-milk cattle of the Shorthorn breed have been almost entirely replaced by the heavy-milking Friesians.

The other great advantage brought by freeze-storage of semen is the ability to export and import the genetic potentialities of different breeds of animal without carrying the diseases and pests of their country of origin—although some viruses may travel successfully with frozen semen and the genes of a particular stock may carry an inherent susceptibility to particular diseases. Most countries have strict quarantine regulations designed to prevent the importing of diseases, and the economic cost of lengthy quarantining of pedigree animals can be prohibitively high. But nowadays, by using frozen semen, the cattle stock of many countries is being improved by bringing in desirable traits from specialized breeds in other parts of the world.

Artificial insemination is not widely used in the breeding of other types of farm animals, so comparatively little work has been done on goat, sheep or horse semen. It might, however, be advantageous to start a programme of AI for pigs. The problem here is the practical one that a boar produces such enormous quantities of ejaculate that it has been found to be unmanageable. There is a certain amount of work currently in progress towards solving this problem. One of the most hopeful suggestions is to prepare pellets of frozen boar semen, a technique which has been tried experimentally with bull-semen though it has not gained widespread acceptance among cattle-breeders. Experiments with freezing fish sperm have been successful. They have enabled the two races of herring, the autumn and spring breeding races, to be crossed for the first time. It seems likely that the growth of fish-farming in the future will eventually encourage further work in this field.

In view of the many years of effort that have had to be put into solving the problems of freezing animal sperm, it is somewhat ironic that human sperm has been shown to be remarkably resistant to the dangers of freezing. Very little trouble has been encountered in preserving human sperm for long periods at low temperatures. It has been formally reported that as many as 1,000 human babies have now been born as a result of insemination with freeze-stored sperm. Most of this work has been done in the USA but there have been a small number of cases in many other countries, including Britain. The ethical and social problems brought up by human insemination with stored sperm are obvious, but a more likely suggestion as to why this work has not been more vigorously pursued has been suggested by Alan Parkes. In the same paper in which he claimed, 'We have abolished time and space in cattle breeding,' he points out that there is 'a marked lack of enthusiasm for abolishing time and space in human reproduction'.

We must now go back to 1952 and Mill Hill and Audrey Smith. After her discovery of the way to freeze human red blood cells, she had been firmly turned back by Alan Parkes to the study of reproductive tissues at low temperatures. He asked her to look into the possibilities of freezing eggs to complement the earlier work on sperm, and the rabbit was chosen as the special animal because it was easily obtainable in their laboratory.

There had been much earlier work on the behaviour of eggs of many species when cooled down below body temperature to levels around 0°C. These had shown clearly that the unfertilized egg was very unstable—many forms of stress including cold would make it start dividing parthenogenetically (i.e. without being fertilized). But after a few stages of cell-division the egg would stop developing and die. Fertilized eggs appeared to be more stable and more able to stand up to handling—some experiments had shown that fertilized eggs could be transferred from the true mother at an early stage and implanted into the womb of a 'foster-mother', where they gave rise to normal offspring. Some successful results had even been achieved with fertilized eggs stored for short periods in normal refrigeration cabinets at temperatures of 10°C. Dr Smith, therefore, did not bother with unfertilized eggs, but concentrated on newly fertilized eggs, which can be obtained reasonably easily from recently mated female rabbits. Her first discovery was that these eggs did not take kindly to being treated with glycerol—like the rabbit sperm they shrank and died from dehydration caused by the strong solution of glycerol surrounding them. It was only by a slow process of steadily increasing the strength of the glycerol solution that they could be persuaded to tolerate the cryoprotectant. But at least she could show that glycerol was not fatally toxic to fertilized eggs if carefully applied.

She then tested the powers of recovery of the glycerol-treated fertilized eggs from various low temperatures, removing the glycerol as they were warmed up. Down to temperatures of minus 20°C the eggs were unstable, but at minus 79 and in liquid air at minus 190°C there was stability. Rapidly cooled eggs never survived, but slowly cooled eggs did show some survival after thawing. However, only six out of six hundred rabbit eggs showed any ability to resume their development after freeze-storing. And all these six died at an early stage. These results were too discouraging to pursue at that time. But Audrey Smith had shown that it was possible to take the fertilized egg down into the low temperatures, thereby arresting its development until it was thawed out, when development could start again. Twenty years later, with improved techniques, her work would be brought to fruition by others.

But in the course of this work she had noticed something interesting among the generally rather depressing results. Certain cells in the covering surrounding the egg appeared to survive the freezing treatment and storage better than the other cells. These tough cells are called granulosa cells and they are essentially part of the tissue of the ovary from which the egg is shed before fertilization can take place.

As well as storing and producing female eggs, the ovary also produces some hormones, the 'chemical messengers' of the body which, though they circulate in minutely small quantities, 'instruct' the large organs to come into action or regulate their activity. The hormones produced by the ovaries play an enormous role in regulating female cyclical activity and all other features of the reproductive mechanism. These subjects were the central interests of Alan Parkes's work. He immediately saw the importance of Audrey Smith's observations about the granulosa cells, and joined her in a series of experiments to see if ovarian tissue could be frozen and stored successfully.

As with all freezing and storing experiments, one of the problems was to find a way of showing whether the frozen cells had fully recovered all their powers when they were thawed out. With ovarian tissue this was comparatively simple. Many earlier workers had shown that if the ovaries were removed from their normal place in an animal and reimplanted in other places in the animal's body they would soon start producing their hormones again. This could easily be judged by noting the stopping of the female cyclical activity when the ovary was removed and then seeing the cycle start again when the ovary was implanted almost anywhere else in the body.

The main problem Audrey Smith and Alan Parkes faced was that if they took whole ovaries from their experimental rats, or even large portions, such as a half or a third, the glycerol seemed to fail to protect them during freezing, and the grafts failed to 'take' when they were reimplanted after thawing. The answer was to cut the ovary up into very small pieces, so that the glycerol penetrated the tissues sufficiently to protect them during freezing. Then they showed that tissue treated by slow cooling survived the journey down to minus 79 degrees and was successful in restarting the rat's reproductive cycle when reimplanted. Fast-frozen tissue did not work. The principle had been firmly established by 1954 that endocrine, i.e. hormone-producing, tissue could be successfully frozen and stored and would return to its normal activity after thawing.

There were, however, two interesting anomalies in their results. First, there was the discovery that quite a large proportion of ovarian cells would stand freezing down to minus 79°C even when unprotected by glycerol. On the other hand, even in the presence of glycerol there was a considerable deterioration in the activity of ovarian tissue in the first few days of storage at minus

79 degrees, though this levelled off and little further damage seemed to be done by longer storage. This deterioration was nothing like so serious when storage was carried out at liquid air temperature, minus 190°C. There was thus a plain indication that for storage of several types of tissue the lower temperatures were more satisfactory—the colder the better—and this was emphasized later when it was shown that sperm were also stored more satisfactorily at lower temperatures, i.e. in liquid nitrogen rather than in solid carbon dioxide. Quite as important was the theoretical evidence these experiments gave that there must be some changes going on in the tissues even at such incredibly low temperatures as minus 79 degrees, and these changes could only be halted by going down to minus 190°C. These changes could not possibly be due to biochemical actions, which were known to be halted far above minus 79 degrees. The damage must therefore be due to some physical cause. Therefore there must be some processes affecting cells which were neither the growth of ice crystals nor the effects of osmotic pressures, and these processes are now, twenty years later, one of the chief puzzles that cryobiologists are trying to solve.

Dr Smith and Dr Parkes took their work further by examining other endocrine tissues. They showed that cells from the testis and adrenal gland of the rat could be successfully frozen and thawed. This opened the possibility that a wide variety of endocrine tissues could be stored at some time in the future. Alan Parkes in particular was fired by the prospect of creating 'banks' of organs and tissues which could be transplanted into patients whose own glands or organs had failed. These were the years when the science of immunology was beginning to take a new course, away from its former concentration on vaccines and immunization, into the elucidation of the body's natural defence mechanisms and the problems of transplantation and rejection. Peter Medawar (later to become Sir Peter Medawar and Director of the Mill Hill laboratories when they became the National Institute for Medical Research) had just completed his first series of experiments which demonstrated the nature of the immune mechanism through the performance and rejection of skin grafts. It was this work which pointed towards the possibility that organs could be transplanted provided the rejection mechanism was overcome. And almost as a by-product of his work, he and Rupert

Billingham had shown that skin could successfully be frozen and stored. The storage of skin in deep-freeze conditions is now a regular procedure at those centres which specialize in plastic surgery and the treatment of extensive burns.

The work of Audrey Smith and Alan Parkes on the storage of endocrine tissue has not yet resulted in great practical applications, but it opened up many prospects which are still being pursued, though not on a large scale. Some immunologically active tissue can be stored and has been used in treating babies born with grave deficiencies in their defensive systems. And a number of experts hope that one day the transplantation of tissues from the pancreas will be a regular treatment for diabetes. This would obviously involve storage of pancreatic tissue.

Audrey Smith herself grew to dislike the work on endocrine tissues, largely because it involved so many operative procedures on laboratory animals. She still says that she tends to 'prefer animals to scientists'. Cryobiology was now an expanding science and other teams were making notable contributions. At this time, for instance, in 1954 a team at the Radiobiological Laboratories at Harwell showed that bone marrow cells could be safely frozen and thawed back to life. This was part of the work on the effects of radiation on the body—for heavy irradiation destroys the marrow cells which are the source of the white cells which defend the body, and the Harwell scientists showed that the marrow could be restored to activity by using stored marrow cells. Dr Smith therefore welcomed an entirely new and much more sweeping development of cryobiology—the freezing of whole animals.

This work was introduced to Mill Hill by the Yugoslav scientist, Dr. R. K. Andjus. Towards the end of the Second World War the library of the University of Belgrade was destroyed in an air raid. Dr Andjus had, therefore, no chance of finding out that science had firmly established that rats inevitably died if their temperature was reduced to 15°C. He reduced rats to a temperature only just above freezing point, about 1°C, after anaesthetizing them, by which time all breathing had ceased and the heart had stopped pumping. He was then able to resuscitate the animals and bring them back to life by a complicated procedure which included placing a red-hot metal spatula on the animal's chest to warm the heart first and get it beating again. He also had to give the rat artificial respiration. He was found doing this work at the Collège de France in Paris by Henry Sloviter, the American member of the Mill Hill team, who had arranged a visit to France as a relief from his work on freezing red blood cells. Sloviter was excited by Andjus's results and in turn he interested Parkes, who invited Andjus to Mill Hill. Sloviter had met Andjus because they were the only two people speaking English in the Collège de France, and this was one of the factors that persuaded the Yugoslav to emigrate yet again. Thus the Mill Hill laboratories became pioneers in the techniques of freezing whole animals.

The next five years produced some extraordinary results, both at Mill Hill and at other laboratories which took up the study of freezing whole animals. It was established that many types of animals could be reduced to a state in which all breathing and heart-beating stopped for periods of more than an hour, and could then be resuscitated with quite a good long-term survival rate. This was done with many common laboratory animals like rats, hamsters and mice, and also with dogs and monkeys. Two American scientists actually succeeded in doing it with a human patient who was suffering from widespread and incurable cancer. This patient was reduced to a deep body temperature as low as 9°C and was resuscitated and lived a further thirty-eight days until she died of her cancer. Audrey Smith froze pregnant golden hamsters and resuscitated them so well that they gave birth to their offspring. It was shown that a small animal could be frozen so hard that it could be supported by its head and tail alone, looking like a solid bar, with weights placed upon it, and still it could be brought 'back to life'. The sight of solid-frozen grey-coloured ears and feet slowly turning back to soft warm pink flesh was often quite moving.

The chief problems were encountered in the process of thawing out the frozen animals, for it was necessary to try and warm them up inside first so that their organs would start working before the skin and outer regions demanded supplies of blood and body liquids. Special diathermy kits were made for the purpose to apply what we now recognize in our kitchens as micro-wave heating. At Mill Hill Jimmy Lovelock made one such machine for Dr Smith.

There were, however, many puzzling features among the host of results achieved. Apparently successfully resuscitated animals were found to have internal injuries, such as perforated stomach ulcers, which could neither be properly accounted for, nor predicted, nor regularly reproduced. Surprisingly, frostbite was not a major problem, though on one occasion a resuscitated animal was found to be suffering very badly from this damage to its ears and tail. It had been shown in the frozen state to a scientific audience and the damage turned out to have been caused by the scientists handling its extremities to assure themselves that it was frozen solid. Eventually it began to be realised that in freezing whole animals the research workers were groping in the dark. There was too much they did not know, too many factors they could not measure, too much complication in an entire body. Much of the interest in this work has disappeared.

Luyet himself had outlined the main difficulty before ever the work was taken up on a large scale. He had experimented with goldfish and shown that if a fish was taken to such a low temperature that it was frozen right through and became brittle so that it could be broken, it did not survive. It was possible to freeze a goldfish so that it became stiff and ice appeared under the skin, yet there was still liquid inside it and this fish would revive immediately on being warmed. Luyet sometimes demonstrated this difference in public and showed his audience how a skilled conjurer could deceive them into thinking that the totally frozen fish (often immersed in liquid oxygen) could be revived whereas he was really reviving the fish that had only been frozen on the outside. Unfortunately Luyet became so good at this demonstration that he appears to have confused some of those who watched him and it has entered the literature that he did, in fact, believe he could revive totally frozen goldfish.

What the research on freezing whole animals could not show satisfactorily was how much liquid remained inside the animals—how much they were supercooled rather than frozen right through. It did, however, lead to two positive results. It showed the way to the use of hypothermia—deep cooling of the body—for surgical operations. It also raised again the fascinating question of the definition of life and death. An animal cooled to just above freezing point has no breathing, no heart beat. To all clinical appearance it is dead, yet it can be revived. Alan Parkes eventually produced this answer: 'Death is the state from which resuscitation of the body as a whole is impossible by currently known means.

At the end of the 1950s the original Mill Hill team had broken up. Alan Parkes took up a professorship at Cambridge, Audrey Smith suffered from ill-health for a year. When she returned to work she needed a new task, and it came in the form of a request by Sir Benjamin Rycroft that she would look into the difficulties of freezing corneas. The cornea is the clear glassy centre-front of the eye, literally the 'window on the world'. It can be damaged by physical injury or by disease so that it loses its clarity. The operation of taking out the damaged cornea and replacing it with a clear one was well known and widely and successfully practised. But the clear corneas normally came from those who had recently died and bequeathed their eyes to medicine. The desirability of setting up a bank to store corneas was obvious. The human cornea had first been frozen in 1954, but although a number of successful transplants were performed with frozen corneas, many problems cropped up with setting up a permanent banking system.

Audrey Smith considered the problem and decided that the likeliest cause of the difficulties was that the cornea itself was rather impermeable to glycerol and other cryoprotectants such as DMSO. This was preventing the extremely thin layer of epithelial (skin) cells at the back of the cornea from getting proper protection. There were also problems with the structure of the cornea itself, especially during thawing, and Dr Smith got the idea that the cornea might be getting waterlogged in the solution as it was being warmed up.

Her collaborator in this work was Dr (now Professor) Michael Ashwood-Smith. He developed a device which was basically a fine hollow needle with apertures some distance back from the point. The needle was put right through the whole eye (after it had been taken from a recently dead donor) so that the apertures were inside the liquid-filled centre of the eye. Through these apertures a solution containing glycerol was introduced into the centre of the eye, and the natural liquid—the aqueous humour—was drawn out. Then the whole eye could be frozen successfully, since the cryoprotectant had been able to penetrate the thin layer of epithelial cells inside the cornea. There were other minor techniques that were also essential to the preservation of eyes, and it was only on the hundred and first experimental rabbit that the system was finally proved successful. But nowadays the world contains a number of eye banks and the transplantation of corneas is the most regularly successful of all transplant operations.

Audrey Smith had one other major 'first' still to come. She was finally successful in freezing and storing cartilage tissue from human joints. Whether this technique will eventually lead to the banking of cartilage and regular operations to replace the damaged cartilage of arthritic and rheumatic joints remains a question that only the future will be able to answer.

During the years when Mill Hill had been struggling with the problems of freezing and thawing whole animals, the science of cryobiology had been rapidly expanding throughout the world. One of the major studies at many other laboratories throughout the 1950s had been the freezing and storage of micro-organisms —bacteria, yeasts, protozoa and others. This work started from a basis of much wider knowledge than the rest of cryobiology, for there had been many studies of the reactions of micro-organisms to low temperatures. Much of this had been in support of the work on the freezing and preservation of food, where obviously it was desirable to know which bacteria could survive or even breed at low temperatures. But in the 1950s, with the advent of cryoprotectants, much progress could be made with the freeze-preservation of organisms which were not naturally resistant to cold. In many laboratories this was a much-needed advance, for many dangerous strains of bacteria and protozoa could only be kept available for study and reference by continuous passaging through generation after generation of laboratory animals. When they could be stored permanently in a frozen state, yet easily resuscitated when they were needed, the laboratory technicians were freed from a great deal of routine, yet potentially hazardous work. The outstanding name among the many cryobiologists who worked in this field of micro-organisms is the American Peter Mazur, of the Biology Division of the Oak Ridge National Laboratory in Tennessee.

Towards the end of the 1960s the main weight of cryobiological effort was turned to efforts to preserve whole organs, with the ultimate object of setting up organ banks which would provide an ever-ready supply of 'spare parts' for patients of the future suffering from failure of heart, kidneys, lungs or liver. Transplant operations at that time seemed to be becoming a real possibility.

The last ten years have shown that the preservation of organs is even more difficult than persuading the body to accept grafted organs from other individuals. Only with the kidneys have the transplant surgeons, the immunologists and the physicians really been successful on such a large scale that transplantation has become the treatment of choice. There seems little reason to doubt that the problems with other organs will eventually be overcome, but no one could really claim that they have yet been conquered. The cryobiologists have not yet succeeded, however, in storing any organ for more than seventy-two hours. They are, nevertheless, still working hard to solve the problems of organ storage.

The American pioneer of cryobiology, Harold T. Meryman, wrote: 'The fact that individual cells can be preserved by freezing carries the implication that tissues, organs and even whole animals can be similarly suspended. Herein lies one of the hazards of cryobiology. Not all cells respond equally well to freezing. In particular not all cells respond equally well to any single freezing procedure. For some species the percentage recovery is very small. Whereas this may not be a major consideration for microbial or tissue cultures, when organised tissues are involved the proportion of living cells necessary to maintain integrity of the structure will be high. When more complex systems such as organs are frozen, where several types of cells and tissues are present and where complex functional relationships controlling the transfer of metabolites and information are involved, techniques which may permit the survival of only a proportion of isolated cells are clearly inadequate. Experimental evidence has as yet offered little encouragement to the proposition that whole mammalian organs may be frozen to stabilizing temperatures with recovery. To project beyond this to the freezing of whole animals is merely visionary. Although no scientific evidence exists that such achievements are impossible, this is no proof of the converse. That visions have in the past come true is no guarantee that all science fiction is but prologue to fact.' He wrote that in 1966. After ten years further work it remains entirely true.

That passage was part of Harold Meryman's introduction to a book covering many facets of cryobiology. Another paragraph of that introduction is equally impressive in its balance and wisdom, and has survived the intervening years better than any frozen organ. 'Although the availability of suspended animation through freezing has made cryobiology a useful tool in many laboratories, it is perhaps the very diversity of its (cryobiology's) applications that helps to perpetuate its state of undevelopment . . . Although the recent literature on cryobiology is vast, it is dominated almost entirely by applications, and although careful measurements may have been reported, they have generally been made on cells or organisms of unique interest to the investigator, producing data which cannot necessarily be extrapolated to other systems investigated by other workers. The student of cryobiology will find himself overwhelmed by quantities of unrelated and often apparently conflicting data, the result of variability in such vital parameters as freezing and thawing rates, storage temperatures, suspending medium, and, inevitably, the experimental material. Even the investigator merely seeking advice regarding a particular application of freezing will find no firm guide lines by which predictions may be made, but only empirical generalities on which to base trial and error experimentation.'

So some powerful cryobiology groups, such as the one at the Clinical Research Centre at Northwick Park near London, which is the direct descendant of the Mill Hill team, are now devoting much of their effort to the basic science of cryobiology. They are trying to answer questions such as, 'Why are cells killed in the freezing process?' and 'How do cryoprotectants work?' Whereas the whole of cryobiology started purely empirically from the chance discovery of how to use glycerol to protect cells during freezing, 'the whole thing is now beginning to seem more rational and intelligible,' in the words of David Pegg, the leader of the Northwick Park Group.

This group, and most other cryobiologists, now attach little importance to the formation of ice-crystals either inside or outside the cells as a cause of damage to the cells through physical disruption. What is now believed is that as ice forms in the solution surrounding the cells the salts become more and more concentrated in the steadily decreasing amount of liquid water. (In the living body sodium chloride, common salt, is indeed the commonest salt, though there are also other sodium salts and potassium salts.) This creates an imbalance of salt concentration across the membrane of the cell wall which draws water out of the cell, thus causing cell shrinkage, in the first stages of freezing, and has exactly the opposite effects during thawing. For during freezing the water outside the cell freezes first, while during thawing the water inside the cell melts last.

It is also likely that during these changes in salt concentrations some of the complex molecules that make up the cell-membrane may be changed or damaged in some way so that the cell becomes leaky; this may show up as extra damage during thawing. The work done by the cryoprotectant is to reduce the damage caused by the 'solution effects', by concentrating at the same rate as the salts are driven out of the freezing water, thus keeping a balance across the cell membrane. The cryoprotectants can themselves cause damage during thawing, as we have seen, if they do not come out of the cells as fast as the water comes in.

In addition to this main 'chemical' problem during freezing and thawing, there are undoubtedly some physical problems too. Cells must suffer 'thermal shock' during the removal or addition of heat, as any material does, as well as 'dilution shock' as they come back to liquid form during thawing. And all the processes involved in freezing and thawing add to the stresses on the cells—they are shaken and stirred about, they may even be centrifuged and whirled around during the removal of cryoprotectant. In addition they necessarily undergo shrinking and stretching as water flows out or flows into them.

Now that more is known about the stresses to which cells are subjected during cooling it is possible to arrange cooling regimes adapted to the particular type of cell. In addition to the slow cooling regimes, there is now a new version of rapid cooling, again invented by Luyet, called 'two-stage cooling'. In this programme the cells are cooled very rapidly to about minus 25°C, then held for some minutes at this temperature, and then cooled rapidly again to minus 196°C. The reasons this can be successful were discovered as recently as 1975. It seems that the cells shrink as they are being held at minus 25 degrees, but because the dehydration takes place at such a low temperature the chemical reaction rates are slower and so less damage is done. When the cells are then taken down to liquid nitrogen temperature they do not freeze internally because the water has been taken out of them.

By using the various different cooling techniques it is now possible, according to the Northwick Park Group, to freeze virtually any type of cell successfully, with one notable exception. The exception is the polymorph granulocyte, or circulating macrophage. This is one of the body's white cells with a very particular role to play—it is full of lysosomes, which can best be described as small bags full of powerful digestive enzymes, and its task is to break up and destroy any clumps of other white cells and the invading bacteria they have killed. It is believed that the lysosome membranes are even more delicate than the membranes of the granulocyte and that the lysosomes rupture under the stresses of freezing, thus killing the cell. The inability to freeze granulocytes is important and particularly frustrating because the basic cause of a number of diseases is the ineffectiveness of granulocytes, and a transfusion of stored granulocytes from donors would appear to hold considerable hope as a treatment.

However, among recent advances attained by more careful consideration of the needs of individual cell types, there has been a very useful improvement in the ability to store platelets. These are the 'first-aid boxes' of the blood system which are rushed to the site of any injury where they produce many of the substances needed to cope with damage.

It is now also possible to freeze and store lymphocytes, the commonest of the white cells of the blood, so that 100% of the cells can be thawed back to full activity. The lymphocytes are the cells which respond to specific invaders—that is to say that only lymphocytes which are specially designed to cope with one particular invader, say the measles virus, will react against it. When the virus enters the body the 'anti-measles' lymphocytes start multiplying and performing their various activities to defeat the invader, while all the other lymphocytes remain unaffected. A recent, and possibly very significant, observation has been made by John Farrant at Northwick Park. He has noted that lymphocytes survive freezing in different proportions according to whether they are reacting against an invader or not. This opens the possibility that scientists may be able to differentiate between one 'family' of lymphocytes and another through their differential response to freezing. An enormously valuable technique this would be, if it proves successful, for at the moment there is no satisfactory way of sorting out the different groups of lymphocytes without committing them to total attack on an invader.

An even more 'rational' approach to the task of matching cooling regimes to the needs of individual types of cells has brought the Northwick Park group an advance which may open up still wider possibilities. They have successfully cooled and stored smooth muscle at a temperature of minus 79°C and have shown that after thawing it retained its power to contract when stimulated. Smooth muscle is the type of muscle which we have deep inside our bodies supplying the control and movement of organs like stomach, gut or womb. It differs from the muscles in our limbs both in its structure and in its control, for smooth muscle is operated automatically by the autonomic nervous system, where limb muscle is under conscious control.

The original observation was made by John Farrant, again, when he started to study the mechanisms of freezing injury in smooth muscle. He began by working out a system which would allow him to tell whether smooth muscle was working properly or not, so that he could tell whether the muscle had survived cooling or not. The work of many physiologists studying smooth muscle could be called on here, and he selected a well-known technique of measuring the amount of contraction of a piece of muscle, immersed in an organ-bath at normal body temperature of 37°C, when stimulated with precise amounts of histamine.

He found that he could get the smooth muscle to recover from cooling in much better shape if he increased the concentration of the cryoprotectant DMSO (dimethylsulphoxide) while the cooling was going on. Other members of the team began to work on the system and found that the acidity of the solution in which the muscle was bathed had considerable effects on its ability to recover. And then they found that reversing the various steps they had taken during the cooling procedure helped the muscle tissue to withstand the strains of rewarming. They now use a medium containing DMSO and PIPES (which stands for piperazine-NN′-bis-2-ethanesulphonate). The concentration of DMSO is increased in five separate steps at 37 degrees, at minus 7 degrees, at minus 14 degrees, at minus 22 degrees and at minus 39 degrees. At each step plenty of time is allowed for the additional DMSO to penetrate the tissue. Cooling is then continued down to minus 79 degrees. In fact, under this regime the muscle is never actually frozen at any time.

For thawing, the process has to be reversed, with the concentration of cryoprotectant reduced in five stages, thus keeping the salt balance the same at all times. The response of the muscle after the process is just as strong as it was before the journey down to low temperatures. There does seem to be somewhat unusual relaxation of the muscle later on, but electron microscope pictures show no difference between cooled and uncooled muscle.

Obviously this system is much more complicated and labour-consuming than any of the freezing methods used for semen or blood cells, so it could only be used if some very special benefit were to be obtained. And the special benefit that they hope to achieve is that this technique may show the way to preserving whole organs at sub-zero temperatures. It is, in the view of this group, the first real sign that we may one day be able to solve the problems of freezing and storing whole organs and thus be able to bank them to await the demands of transplant patients.

One other very recent advance in cryobiology makes a satisfactory end to this chapter, because its practical application will make it one of the most valuable of the contributions of low-temperature science. It is the solution to the problem of how to freeze and store fertilized animal eggs—in fact, the storage of embryos in deep-freeze.

One of the pioneers in the 'rational' approach to freezing has been Peter Mazur at the Oak Ridge Laboratory in Tennessee, whose work went much of the way to establishing that 'solution effects' were one of the major causes of damage during freezing. And it was in Mazur's laboratories that David Whittingham, a young Englishman, was the first to freeze, store and recover an embryo. This was in 1972, twenty years after Audrey Smith had shown that a very small proportion of fertilized rabbit eggs could be persuaded at least to continue dividing for a short time after being frozen and thawed. Dr Smith had certainly not got any of the fertilized eggs to develop into normal animals, but Whittingham was able to implant his frozen and thawed mouse eggs into the reproductive systems of 'foster' mothers and get normal live young mice from a reasonably high proportion of his experimental animals.

The fertilized egg contains the material necessary for producing an entire individual. Scientists outside the field of cryobiology had shown it was perfectly possible to obtain newly fertilized eggs from the females of many species of animals. They could obtain the eggs before they had started dividing, or when they had divided into two cells, or four cells or eight cells or at the later stages when the growing embryo is called a morula or a blastocyst.

They could then transfer these embryos into other females in which they would develop and eventually become fully-formed, naturally born babies. The foster mother bore no relationship to the infant and it had been shown to be perfectly possible for an animal of one breed to gestate and deliver an animal of another breed—for instance a Hereford calf could be implanted into a Friesian mother and could develop perfectly normally as a Hereford.

There was therefore a fully developed technique for showing in the clearest way that embryos had survived freezing and storage. In an extensive series of experiments David Whittingham showed that by very slow cooling—as little as one-third of a degree per minute, by slow thawing at four degrees per minute, and by careful manipulation of the cryoprotectant—adding DMSO in stages before starting cooling and removal after thawing—he could get very good survival of early embryos of the mouse. He obtained 70% survival of two-cell embryos, and then 65% of his foster mothers became pregnant and 40% produced normal live baby mice. The embryos had been stored at liquid nitrogen temperature for eight days. There had, however, been poorer results when using the later stage of embryo called the blastocyst.

These dramatic results were almost immediately confirmed by Ian Wilmut who was working on very similar lines at the Agricultural Research Council's Unit of Reproductive Physiology at Cambridge. This rather lengthy title is the modern name of the laboratory in which Chris Polge, with Leslie Rowson, had pioneered the breeding of calves using frozen bull sperm, and both these men are still working there. They took up the lead that Whittingham had provided, and by using even slower rates of cooling they and their team successfully preserved cow embryos within a year. The first calf born after storage for several days in liquid nitrogen was successfully delivered in 1973. It was called Frosty, the name given to Polge's first calf from frozen semen. The following year, 1974, this team got the first successful lambs that had been frozen as embryos.

Meanwhile David Whittingham had returned to Britain, where he now works in a Medical Research Council Unit that is part of University College, London. He rapidly improved the techniques for storing mouse embryos until he could achieve 100% survival rates for embryos stored for as long as eight months in liquid nitrogen. He was also successful during 1974 and 1975 in freezing, storing, and later bringing to normal delivery, both rat and rabbit embryos.

When these different teams of scientists compare their results, however, strange variations are seen. Although mouse embryos can be frozen and stored when they have reached the eight-cell stage or the even later blastocyst stage, much the better results are achieved by taking very early embryos in the one-cell and two-cell stages of development. Rat and rabbit embryos survive best when frozen as four-cell, eight-cell or morula stages. But no one has succeeded in persuading early-stage cow and sheep embryos to survive. The successes with these larger animals have all been achieved by freezing the embryos as blastocysts and morulae. Again, sheep have proved much tougher embryos than cows—the success rate with cattle is still very low but Chris Polge has been able to get 50% survival with sheep almost from the start. No one knows why such slow cooling rates are required for embryos, and even less can anyone suggest why they need such slow thawing rates.

One of the principal technical problems faced by cryobiologists working on embryo storage is that of getting the foster mother into the correct condition of receptiveness for the implantation of the thawed embryo. There are reasons to believe that only a few hours' difference between the stage of the foster mother in her reproductive cycle, and the time at which an embryo of the size implanted by the scientist would normally be present, may render the whole experiment abortive. In mice and rats the female is made 'pseudo-pregnant' by mating her to a sterilized male, so that the progress of her reproductive cycle can be timed exactly. But when the foster mother is a cow or sheep this cannot be done and the scientist has to try and coincide his implantation with the animal's natural oestrus cycle. It is very possible, therefore, that inefficiency in achieving pregnancy and full development of the embryo—which in this case is no more than a system for measuring the results of freezing and storing—may be masking the true rates of success and failure in the strictly cryobiological parts of the experiments.

With large farm animals, such as cows, the significance of successful embryo storage is commercial. The technique of ova-transfer has already found a place in the world of pedigree cattle-breeding. An expensive pedigree cow is made to 'superovulate' by a dose of fertility-drug. She is then artificially inseminated with semen from an expensive bull. An unusually large number of fertilized eggs is thus achieved—far more embryos than the mother could gestate herself. The eggs are removed from the true mother and implanted into less expensive foster mothers. Thus a considerable number of valuable pure-bred pedigree calves can be produced. The problems in this system are two: firstly, expensive surgical operations may have to be performed by vets to obtain the fertilized eggs and implant them, involving some danger to the pedigree mother; and secondly, it is difficult to get all the prospective foster mothers into the correct stage of their reproductive cycle at the same time. Embryo storage would obviously overcome this second difficulty. A straight commercial equation can then be applied, comparing the genetic gain from setting up embryo-storage banks against the economic cost. The greater profit in storing cattle and sheep embryos seems likely to come in the import-export market. Embryo storage makes it possible to carry an entire herd of pedigree cattle from one country to another in small glass ampoules in a Dewar flask of liquid nitrogen. The cattle can be foster-mothered by any cows in the importing country without interfering with the known genetic composition of the imported animals.

With smaller animals such as mice, the commercial benefits of embryo storage are not so immediately obvious, but the possibilities are even more dramatic. David Whittingham is now carrying out experiments to try and prove the safety and practicability of banking entire breeds of animals.

The mouse is a successful pest because it multiplies so fast. For the same reason it is a favourite laboratory animal. Strains of mice can be bred until they are genetically so pure that each individual is virtually identical with every other individual—these strains are used in immunological and transplant research because the animals will not reject grafts from others of the same strain. Or, as soon as a naturally occurring mutation crops up, a whole breed of mice can be formed possessing this mutation, enabling scientists to study genetics and the effects of single genes in a population. Some strains of mice are particularly susceptible to certain diseases such as specific types of cancer; other strains are definitely resistant to the same disease. Experiments on hybrids or chimaeras formed from these two strains throw light on the disease. But when a scientist has finished his experiment involving a particular breed of mouse the breed has still to be kept going, although the researcher no longer uses the mice. Others may wish to check his results or develop his findings further. The preservation of the huge numbers of known breeds and strains of mice becomes a considerable burden on laboratory budgets, space and manpower.

David Whittingham's proposal is simply to store these mice as embryos. He has shown in his own experiments that embryos can be stored successfully for at least two years in glass ampoules surrounded by liquid nitrogen. He has developed a machine system which performs automatically the tedious task of cooling at whatever rates are shown to be most satisfactory. The embryos can be recovered in high proportions and any breed of mice can be used as foster mothers without changing the genetic make-up of the embryos they bring to full-term.

The banking of embryos would enable scientists to perform a considerable number of experiments on the development of the embryos that are not at the moment possible because enough embryos at any particular stage of development cannot be obtained. But this is hardly as dramatic as the concept of an entire breed of animals kept, until they are needed again, as invisible small embryos in a little glass ampoule.

Eventually some of the world's larger conservation problems could find their solution in embryo banks. Such banks would enable us to keep breeds of cattle or other productive animals which are for the moment uneconomic but which might contain genetic material, such as disease or radiation resistance, that we will need in the future.

7 Superconductivity - Technology of the Future

Kammerlingh Onnes achieved the liquefaction of helium in his laboratories at Leiden University in 1908. Today, over seventy years later, we are just entering an era of a new technology which has been made available only by our ability to manufacture liquid helium and by our ability to reach the supercold temperatures which are necessary for helium to become liquid. The new technology is based on the greatest discovery of Kammerlingh Onnes's long and distinguished scientific career, the discovery of superconductivity, undoubtedly one of the most significant discoveries of our century.

Naturally enough, once they had discovered how to liquefy helium, the Dutch team studied the properties of the new liquid itself. They found that liquid helium was very light—eight times less dense than water—and that the surface tension of the liquid was also very low. But they were if anything even more interested in studying the properties of other materials at very low temperatures, and they could do this by using liquid helium to cool any other substance to its own temperature of less than five degrees above absolute zero. Indeed, since they could cool liquid helium to about one degree above absolute zero, they could also cool anything else to around this temperature. Their work was, of course, the pursuit of pure science, the study of material behaviour at superlow temperatures. And they had a great theoretical incentive to this work because the Third Law of Thermodynamics had then recently been discovered, implying certain conclusions about absolute zero and what would happen as this ultimate temperature was approached.

The Third Law of Thermodynamics was formulated by Walter Nernst in Berlin in 1906. He had no idea at the time that it was the Third Law of Thermodynamics. He simply made a rather bold assumption about the amounts of energy available in chemicals for use in reacting with other chemicals. Within ten years his assumption was recognized to be a Law of Thermodynamics of as much significance as the two previously recognized major laws. During this period the German chemical industry had used Nernst's assumption as the basis on which were successfully built giant plants which produced much of the highly successful explosives for German armaments in the First World War.

But the theoretical importance of Nernst's Law was much greater than the practical significance of producing armaments, for it made the first scientific statements about the state of matter at or near absolute zero and opened up the theoretical side of all the studies of very low temperatures.

The Third Law of Thermodynamics appears to say two simple things about absolute zero. First, it says that while scientists may lower temperatures again and again and approach closer and closer to zero, the condition can never be finally attained. Secondly, it says that at absolute zero the entropy (that difficult factor in the Second Law of Thermodynamics) becomes zero. When we write that a law 'says' something, it is important to remember that the law itself is stated in terms of a mathematical equation which correctly expresses the measurements made in experiments conducted with temperatures, pressures and other factors at levels that can be attained in the laboratory. The law then predicts what will happen if scientists change these temperatures and pressures. In some cases the mathematical formula predicts what will happen if one of the factors is reduced to zero, and this is the case with the Third Law of Thermodynamics—the mathematics predict that the temperature can never be reduced to zero.

Both the simple statements made by the Third Law about absolute zero have had to be qualified somewhat since Nernst's first bold proposal, but the significance in practical terms remains the same. The first statement, that absolute zero can never be reached, is the more dramatic but the less important. It is best regarded as meaning that we must abandon the simple idea of temperature as something which can always be divided into equal steps or degrees. Just as we can always imagine a higher temperature being reached, because we can always imagine making something a little hotter, so there is always the possibility of making the coldest thing a little colder. It is scientifically more useful to regard an object at 100 degrees as being ten times hotter than an object at 10 degrees, rather than thinking of the first object as being 90 degrees hotter. Correspondingly, an object with a temperature only $\frac{1}{1,000,000}$ °C above absolute zero is only ten times colder than an object at a temperature $\frac{1}{100,000}$ °C above absolute zero. (And these apparently very small temperatures are quite attainable nowadays.)

The second conclusion from Nernst's Law was, however, more important. Strangely enough there had been no scientific interest whatsoever in the likely behaviour of matter as absolute zero was approached, although as we have seen, the possibility of there being such a concept as absolute zero had been known for more than a century. The Third Law directed attention to the likelihood of there being interesting phenomena at very low temperatures and therefore pointed the way to our modern science of cryogenics. The Law also caused much surprise by its prophecy that it would be the entropy that tended to disappear as absolute zero was approached. On the face of things it would have seemed more likely that the energy would disappear.

The chief practical problem facing Kammerlingh Onnes in investigating matter at low temperatures was to design experiments which could actually be immersed in the liquid helium in the centre of the complicated liquefaction apparatus. One possible answer to this problem seemed likely to be the study of the electrical resistance of materials immersed in liquid helium. This had the advantage that it could be applied to the questions raised by the Third Law of Thermodynamics, which seemed to imply that resistance would grow gradually smaller until it finally became zero at absolute zero. However, Nernst's Law could also be interpreted as implying that all the electrons in matter would become absolutely fixed to their appropriate atoms in a state of perfect order at absolute zero, and therefore the resistance would become infinitely large.

Onnes started work on measuring the resistance of platinum in liquid helium. He found neither of the predictions about resistance to be true—it apparently neither increased nor decreased as temperature was lowered. But it was noted that the measurements varied from sample to sample and Onnes guessed—correctly—that the crucial factor must be impurities in the metal. And he guessed—incorrectly—that with pure materials resistance would become zero at absolute zero, or a few degrees above, following a steady fall of resistance as temperature lowered.

What had to be done, then, was to obtain the purest possible samples of a metal. Onnes at first tried to get extremely pure gold, but he could not rid it of the final minute proportions of impurity. So he switched to mercury, a metal and a good conductor of electricity, which is liquid at room temperatures and can, therefore, be distilled over and over again with ease, until an exceptionally high degree of purity is reached.

It was the Royal Netherlands Academy that heard of the great discovery on 28 April 1911. But none of its members, nor Kammerlingh Onnes himself, realized what he had done. His paper showed that mercury, as pure a sample as man could make, still had an easily measurable electrical resistance at the temperature of liquid hydrogen; still had a resistance at the boiling point of liquid helium, although it was only just measurable; but had no resistance at all at some even lower temperature which was not very clearly defined. Onnes pointed out that this was exactly in agreement with his predictions but, wisely for his reputation, left in the safeguard that these were preliminary results and promised more accurate measurements later.

One month later the more accurate measurements were provided. True, there was a measurable resistance in mercury at the boiling point of helium—4·2 degrees above absolute zero. But just below that temperature the resistance of the metal vanished totally, in a rush, instead of declining to zero gently as the temperature was steadily lowered. Onnes merely noted this oddity, and then there was seven months' scientific silence on the subject.

It seems that many of the most important scientific discoveries are reached because a great scientist, a man of peculiarly acute imagination, has spotted an irregularity in the otherwise apparently regular behaviour of nature, and has realized that this irregularity is a clue, leading to some fundamentally different phenomenon. Onnes' next communication was titled 'On the sudden change in the rate at which the resistance of mercury disappears'. It did little more than confirm the result he had published before and provide a precise temperature at which the phenomenon occurred—a precision within two-hundredths of a degree.

Then there was a further silence of almost a year—the whole of 1912. But in the first five months of 1913 there appeared no fewer than four papers from Onnes showing what he had been doing throughout the previous year. He confirmed that the disappearance of the electrical resistance in mercury was not the phenomenon he had originally predicted, the steady reduction of resistance to zero as absolute zero was approached. It was something quite different, the sudden disappearance of all resistance at a point just below the temperature at which helium liquefies. Exactly the same phenomenon occurs with tin and lead, he found, though the temperature at which the disappearance of resistance occurs is different with each metal.

And in the second of his 1913 papers Onnes acknowledges the existence of a new phenomenon—he names it superconductivity. Superconductivity is the state in which an electric current will flow apparently for ever and undiminished in a material because all electrical resistance has disappeared.

Kammerlingh Onnes himself had great difficulty in recognizing the importance of the discovery he had made, for not unnaturally, he tended to regard superconductivity as a special case of ordinary electrical conductivity, the subject on which he had been working. It is always difficult to realize that one has broken out of the paradigm that rules one's own professional subject or way of life.

Nevertheless, the community of science recognized the importance of Onnes' work and awarded him the Nobel Prize in 1913 and many other distinctions. Superconductivity was not the end of his work—some of his later researches were to lead to equally important progress—and it is perhaps symbolic of the stature he himself gave to his discovery that he would demonstrate superconductivity by a pretty little gimmick. This display depends on the fact that electricity and magnetism are essentially the same force, so that if electric current flows in the wires of a metal coil it creates a magnetic field, while if the magnetic field round a magnet is moved with respect to a conducting wire then an electric current is made to flow in the conductor. Onnes put a current into a coil of lead wire so that a magnetic field was created which could be detected by a compass needle; he then cooled the wire until it became superconductive and cut off the electric current. Because the wire was superconductive, with no resistance, the current continued to flow inside it, despite the fact that the power source had been cut off, and so a compass needle outside the apparatus continued to detect the magnetic field caused by the current in the wire for hours—usually until the supply of liquid helium failed and the wire warmed up to destroy the superconductivity.

A different version of this demonstration, also practised by Onnes, was to put a simple lead ring in the liquid helium and then energize a large magnet outside the cryostat. When the outside magnet was taken away, the magnetic field which had been produced in the lead ring would remain there because the metal was superconducting—any alteration in its immediate magnetic field would be immediately compensated by the induction of a current in the ring which would recreate the magnetic field. A simple compass needle outside the cryostat would then demonstrate the field inside by being deflected.

Slightly more complicated demonstrations are common nowadays, since scientists have become more accustomed to handling liquid helium and can contain it at its low temperature in apparatus made entirely of glass. This makes it possible for us to see 'levitation' in which pieces of metal or magnets are held up in liquid helium without any visible support. It is easy to induce magnetic fields or electric currents in a couple of small lead rings in a bath of liquid helium. Then a lead ball is lowered into the space between the lead rings, without touching either of them, and it will stay floating above the two rings. What has happened is that, as the ball descends through the liquid helium towards the rings, its movement in the magnetic field generated by the magnetism in the rings induces electric currents on the ball's surface. These currents generate a magnetic field in the ball, and finally the mutual repulsion of the magnetic field in the ball and the rings equals the force exerted by gravity on the ball, and it stays positioned just above and away from the rings, apparently floating.

A similar effect can be shown if a lead dish is immersed in liquid helium and a small bar magnet is lowered on a chain towards it. The magnetism in a bar induces electric currents and hence a magnetic field in the superconducting lead dish until eventually these fields become strong enough to repel the bar magnet. The field in the lead dish remains because it is superconducting, and the magnetism in the bar magnet is permanent. The magnet 'floats' above the dish with the chain that has lowered it hanging in a slack loop, so that it can be clearly seen that the magnet is not supported by anything but the repulsion induced in the lead dish. (I have used this demonstration myself on television.) But these are no more than laboratory demonstrations of an interesting phenomenon. There was so little interest in 1913 that superconductivity was not demonstrated in England till after Onnes's death. In 1932 his collaborators Keesom and Flim were persuaded to show it at the Royal Institution, and the Leiden laboratory still treasures the story of this expedition. There were no scheduled airline flights from Holland to England at that time, and Keesom, who was distinctly stout, had great difficulty in getting into the sports plane they had to hire for the journey to England with the precious liquid helium. They also had to design a special cryostat, for they feared that during the flight the air in the head of the cryostat would turn to snow and jam the head. During the flight they hit an airpocket and the new head nearly came off. Then, when they got to London with their precious cargo and demonstrated the superconductivity by using the lead-ring method, the audience simply would not believe that the compass needle was deflected by the magnetic fields 'trapped' in the superconducting lead ring. They thought it was a trick and that some magnet was concealed in the cryostat. Flim and Keesom had to empty the cryostat and demonstrate that there was nothing hidden in it except the lead ring!

It is ironical that these simple early demonstrations of superconductivity themselves led to the main reason why superconductivity seemed likely to remain no more than a laboratory phenomenon, or at least no more than a matter of interest to research physicists working out the final variations in the behaviour of materials under extreme conditions. Kammerlingh Onnes himself realized, soon after he had grasped the nature of his own discovery, that superconductivity offered a way of building very small but extremely powerful magnets. All really powerful magnets are made by winding coils of copper wire round iron cores, and then putting an electric current into the coils of wire to generate a powerful magnetic field in the iron. Because of the resistance of copper to the flow of electricity the wire heats up, and in all really large magnets the problems are caused by the necessity to supply and circulate huge quantities of water to remove the heat generated in the wire coils. In a superconductive coil, however, there should be no heat generated, because there is no resistance; cooling would be confined to bringing the wire down to something like the temperature of liquid helium and keeping it there. Theoretically, it seemed possible that any amount of current could be driven through such a wire.

The disappointments came very early in the programme. It turned out that the moment any substantial magnetic field was created by putting current through a coil of superconducting metal, the superconductivity disappeared, the system simply broke down and the metal started to behave like an ordinary conductor. Likewise, the very early stages of the experiments showed that if more than a certain amount of current was put through a superconducting wire, the superconductivity vanished again. It appeared that superconductivity could only be achieved when the temperature was very low, when the magnetic field was very low, and when the current was very little. So it seemed there could be no practical applications of the new discovery in any useful machine.

Although a number of other laboratories in the world took up research into superconductivity in the late twenties and early thirties of this century, the work was very much 'pure science'. It was shown that tin, indium, gallium and thallium could also be superconductors as well as the original lead and mercury of Onnes's first observations. All these, however, tend to be soft metals with low melting-points. In Berlin, Meissner showed that some hard metals with high melting-points, such as titanium, tantalum and niobium, could also be superconductors. And as the ability to reach temperatures even lower than that of liquid helium spread round the laboratories, and when it became not uncommon to reach as little as one degree above absolute zero, it was demonstrated that still more metals, such as aluminium, cadmium and zinc, could superconduct.

It was from Meissner's laboratory in Berlin that the really disturbing result came. They showed that if a metal sphere was placed in a magnetic field and then cooled to the superconductive condition, the result was exactly the same as if the sphere was first made superconductive and then placed in a magnetic field. This meant that when the sphere went superconductive it actually expelled the magnetic field. Yet, remember, it had also been shown that a high enough magnetic field would completely destroy superconductivity. The theoretical importance was enormous.

Meissner's discoveries, made in 1933, were quickly confirmed by other laboratories now rising in the international field of cryogenics, notably the Clarendon Laboratory at Oxford, and the Russian laboratory at Kharkov. By 1935 it was clear that some alloys of metals could also be made superconductive, though these were not common materials; they were alloys of the rare metal niobium or unusual mixtures such as lead-bismuth. In some cases these alloys showed remarkably high 'critical fields'—which means that although they still needed very low temperatures to become superconductive, they would remain superconductive in comparatively high magnetic fields.

Then it emerged that the superconductive alloys behaved rather differently from the pure metals in the phenomenon of critical magnetic fields. Whereas a pure metal superconductor would have its superconductivity extinguished in an instant by a rising magnetic field, as if the magnetic field had broken through some surface barrier and flooded into the body of the metal, 'quenching' the superconductivity, with the alloys it became clear that the external magnetic field would slowly and partially penetrate the specimen, leaving a superconductive region at the centre which was only slowly quenched as the field rose. Research now began to turn towards finding the properties of the materials that allowed superconductivity to behave so differently. And in particular there began a line of research into very thin wires, where the depth of penetration of the external field was of the same order of magnitude as the diameter of the wire; this would have great repercussions many years later.

Right from the start these experiments showed that the 'ordinary' superconductivity discovered by Onnes was confined to the surface of the materials involved. The superconducting electrical currents flowed only in a very thin layer on the outside of the material. But to account for the behaviour of those alloys where superconductivity decayed at measurable speed as the outside magnetic field partially penetrated the material, there came the ideas of tubes or filaments alternatively of superconducting and normal material existing side by side in the samples under study.

Although at that time (in the 1930s) these ideas were mostly theoretical, inspired by the search for a satisfactory theory of superconductivity, some of them, such as the concept of tubes of superconducting material inside a sample, later came to attain practical importance. The work on very thin specimens also gave rise to the first hints that the availability of paths open to the passage of electrons might provide some explanation of the whole phenomenon of superconductivity.

But towards the end of the 1930s there was a perceptible slowing down, even a near halt, in superconductive research. In Russia Shubnikov, leader of the Kharkov group, disappeared in the Stalinist purges. In Holland, Onnes's successors, such as Gorter and Casimir, were struggling with the intractable problems of theory, but the momentum provided by the great founder of the Leiden laboratory was naturally running down. Scientists in Britain were beginning to prepare themselves for their wartime work, while in the 'purest' physics laboratories attention was drawn away towards the production of even lower temperatures and the extraordinary phenomenon of 'superfluidity' which helium can demonstrate.

However, one interesting perspective was opened by the work of the 1930s and has never been fully exploited in any major technical device. This is the ability to concentrate a magnetic field, or even to pump magnetic flux into an area of storage that is provided by the phenomenon of a superconductor actually expelling a magnetic field.

It was the development of a small and simple machine for providing liquid helium—a cooling device that virtually any laboratory could afford to buy and run—that revived research into superconductivity in the years following the Second World War. The machine was invented by Professor Collins and manufactured by Arthur D. Little in Massachusetts. It wrought such a revolution that cryogenics research men refer to the era before about 1950 as 'BC—before Collins'.

It is always disturbing to contemplate the number of useful advances in technology brought about by the destructive demands of war. But it was undoubtedly the development of wartime devices which led to the vastly increased production of helium in the USA and its ready availability for scientific research after the war was finished. Professor Collins had been commissioned to work in the field of refrigeration as part of the defence effort. And so in the late 1940s his helium-liquefying machine became available to laboratories all over the world, and many establishments took advantage of this availability to start small programmes of low-temperature research. It was no longer necessary to have enormous industrial-style facilities on the Leiden pattern, nor to have a large team of top-class scientists devoting the main drive of their research to cryogenics, as had been the case in the 1930s in Oxford, in order to produce scientifically valuable results in superconductivity or the properties of materials at temperatures near absolute zero.

The Collins helium-liquefier was, in principle, the simplest form of cooling machine, an expansion engine in which the gas is cooled by making it do work to drive a piston. Its intellectual ancestry lay in the work of Simon's pre-war Oxford group and in the brilliant engineering of the Russian Kapitza, who had worked under Rutherford at Cambridge. By careful design and clever engineering Collins had produced an expansion engine which worked regularly and reliably at a temperature only ten degrees above absolute zero—i.e. at a temperature which could be attained by cooling with liquid hydrogen. The machine could operate steadily to provide litre quantities of liquid helium in virtually any laboratory.

The appearance of the Collins helium-liquefier led to what even a stolid textbook calls 'a spectacular growth' of cryogenics as an academic study in physics laboratories all over the world. Judging only by scientific publications on the subject, there was an annual growth of between 10% and 15% in the amount of work on the subject in the 1950s and 1960s. There are now known to be at least 35 elements and 1,000 different materials and alloys which can show superconductivity under the right conditions.

But the picture that emerged from much of this work was confused—indeed, much still remains confusing. It became clear that there were two types of superconductors: the original type, discovered by Onnes, in which the superconductivity disappeared, or was quenched, by fairly small magnetic fields and which would not carry high currents—these were mostly very pure samples of the metallic elements and were called Type I superconductors; and the alloys, called Type II superconductors, which seemed to allow some penetration of the external magnetic field before their superconductivity was quenched. But the distinctions were not clear. The element bismuth would only superconduct if it was under pressure, but then it showed up as a Type II. Manganese would only superconduct if it was perfectly pure. The element niobium, however, was a straightforward Type II superconductor. Many alloys and other materials turned out to be somewhere in between the two Types.

There was also confusion about the theory of superconductivity. Onnes's successors at Leiden, Gorter and Casimir, produced a 'two-fluid' theory even before the war which accounted for some of the measured facts of superconductivity. The brothers F. and H. London, working with Simon's group at the Clarendon Laboratory at Oxford, pressed the matter further, basing their explanations on the behaviour of magnetic fields in superconductivity. When the Russians came back into the field after the war, Ginsburg, Landau, Abrikosov and Gorkov produced an even more satisfactory theory based on the concept of the orderliness of molecules and atoms at very low temperatures. Their work is still called the GLAG theory, after their initials.

The most important theoretical advance came in the USA in 1957 with the publication of the Bardeen, Cooper and Schrieffer theory. This has been modified somewhat over the years; it does not explain the difference between Type I and Type II superconductors satisfactorily; it does not predict which materials will fall into which types; it does not predict which way to look for better superconductors; most scientists feel that it will need further modification; but it is still the best theory we have got for superconductivity and most workers in the field accept it, even if they hope it can be improved. The great advantage of the BCS theory is that it provides some real physical explanation for how superconductivity can happen.

When considering the flow of a normal electric current through a metal we think in terms of the movement of individual electrons through the material. Resistance is the holding-up of the passage of these electrons by any form or irregularity in the ordering of the atoms that go to make up the regular structure of the material —a structure that we normally think of as consisting of many small crystals. Irregularities come at the boundaries of one crystal meeting another, or wherever there are the odd shapes and disruptive sizes of atoms or impurities upsetting the regular structure. As the electrons of the current are stopped or slowed by these irregularities or the variations in electrical attraction and repulsion which they exert, the electrons give up the energy provided by the driving voltage and this energy heats up the conductor.

The Bardeen-Cooper-Schrieffer theory explains superconductivity, not only by the extraordinary amount of orderliness in the material at very low temperatures, but also by allowing for an orderliness in the electrons of the current where new forces can come into play because of the low temperature. It pictures superconducting electrons as travelling in pairs, with each member of a pair spinning in opposite directions. If the conditions are right for such a pair of electrons to be 'in tune' with the vibrations of the whole structure of the material (the name of these vibrations is 'phonons') then the pair of electrons can proceed through the material virtually without resistance. Furthermore, we can think of the pair of electrons as helping each other to bypass the resistance offered by atoms of impurities. The BCS theory is by no means commanding nor all-embracing, nor easy to understand. But it does provide a workable picture—a model—which helps scientists to progress in their practical understanding of the phenomenon of superconductivity.

This theory was only just beginning to hold the stage when the most important advance in superconductivity was made in 1961. This was the discovery of what are called 'high-field superconductors', which has at last made superconductivity a technology, rather than a scientific curiosity.

The break-through came as a result of a conversation between two young scientists at the University of Chicago with Enrico Fermi, the Italian-born nuclear physicist, who had created the world's first controlled nuclear chain reaction at that university during the wartime atomic-bomb programme. The two young men were Bernd Matthias and John Hulme, who were working at that time on advanced dielectrics and were studying in particular the not very common compound, barium titanate. Fermi advised them that superconductivity was the most important subject in low-temperature physics. And the two young men took his words to heart. John Hulme was shortly afterwards recruited by Westinghouse, one of the USA's largest manufacturers of heavy electrical equipment, to research into superconductivity. Following the lines of thought started during the barium titanate work, he investigated even rarer compounds, the A15 compounds, which are specified by their unusual crystal structure. He showed that one of them, a compound of vanadium and silicon (V_3Si), was superconducting. Not only that, but he found that the critical temperature, the temperature at which the compound turned superconducting was 17°K. Seventeen degrees above absolute zero was by far the highest temperature at which a material had been found to go superconductive—it is a temperature nearly up to the mark at which hydrogen gas turns into liquid, and therefore offers the possibility that one might find materials in which one could achieve superconductivity without the problems of getting liquid helium.

Hulme told Matthias of his discovery. Matthias, now at Bell Telephone Laboratories, immediately started a programme of development. He soon found several other similar compounds of the A15 type which would become superconductive at comparatively high temperatures, and in particular he discovered that niobium-tin (Nb_3Sn) had an even higher critical temperature—eighteen degrees above absolute zero.

Bernd Matthias is still searching for superconductors at the time of my writing this book, almost twenty years after his first important discovery in the field. He has found many substances which have this extraordinary capacity for losing all electrical resistance at very low temperatures—indeed, there are known to be more than a thousand superconducting materials. But the important discoveries made by Matthias and others are of those which go on superconducting at high temperatures—if one can think of fifteen or eighteen degrees above absolute zero as a high temperature. Matthias in his public lectures nowadays admits to having certain 'alchemical' rules to guide him in his search—he has found purely by experience that certain numbers and arrangements of electrons seem more likely to produce high-temperature superconductors. But the start of a true technology of superconduction depended also on another major discovery.

This next discovery was made by one of Matthias's colleagues at Bell Telephone Laboratories—Kunzler. He showed in the period around 1960 that several of the high-temperature superconductors were also 'hard'—they resisted invasion by the lines of force of an external magnetic field and remained superconducting even when in large magnetic fields. The most important of them all was niobium-tin—the very alloy which showed the highest critical temperature. When Kunzler tried out niobium-tin in magnetic fields he found that even at 100,000 gauss (a unit of magnetic induction) he could not stop his specimen from acting superconductively. What we now call the 'hard' Type II superconductors had been found—materials which would remain superconducting while carrying currents of a size practical electrical engineers would be interested in, and while working in magnetic fields of the strength produced by industrial-sized magnets. And the best of these materials would be superconducting at temperatures well above the temperature of liquid helium—that is to say, at temperatures which workaday engineers could easily and regularly produce and maintain in the 1960s.

The original dreams of Kammerlingh Onnes and the Oxford pioneers were now becoming possible. One could envisage superconducting electric magnets providing very large fields while consuming virtually no electric current or power, and not needing huge flows of water to cool them down. One could visualize power engineering—electrical power-stations and transmission cables cooled to liquid helium temperatures and working without any electrical resistance or heating difficulties.

The great problem still needing to be solved before these dreams became reality was the mundane obstacle of making actual pieces of working equipment from these magical 'hard' superconductor materials. Magnets—even superconducting ones—need miles of wire which has to be wound in complicated patterns to produce an intense magnetic field within the core area. Niobium-tin is a particularly problematical material to deal with. It is difficult to make as an alloy, and when made it is hard and brittle, very unlike the soft and ductile copper or aluminium that electrical engineers are accustomed to dealing with. All through the last months of 1960 and the beginning of 1961 John Hulme's team at Westinghouse and Matthias's and Kunzler's colleagues at Bell struggled with this problem. Bell Labs succeeded in making niobium-tin wire but only with great difficulty, and presumably at high cost. Westinghouse drove for a different solution and made wires of niobium-zinc and titanium alloys. These alloys had a lower critical temperature than niobium-tin—that is to say they would lose their superconductivity more quickly if the temperature were raised—but they were easier to work as engineering materials and could be drawn into workable wires by pulling through a fine die. Other big manufacturing companies were now joining what looked to be a rapidly developing field, and Westinghouse have a long-standing dispute with North American Aviation as to which company's research team first made a practicable length of superconducting wire.

But by the end of 1961 the first workable superconducting magnets had been built and been shown capable of producing fields of the strength of 60,000 gauss. More than fifteen years have passed since then; there is still no power-station working superconductively, nor any commercial transmission cables working at liquid helium temperatures. The first major installation of superconducting magnets is only now being installed, and superconducting electric motors are still only at the stage of being tested experimentally to see if they can stand up to the wear and tear and dirt of real working conditions. But this time-lag is exactly what one would expect from a study of the introduction of other major new technologies in the past. The gap between research laboratory and daily industrial use seems to be of the order of twenty to thirty years, unless there is the pressure of military or wartime needs to speed up the introduction of the technology.

The reasons for this normal time-lag are threefold. First, there is the transition in engineering terms between the successful performance of a small device in clean, controlled, laboratory conditions and the requirement for continuous reliable service of a full-scale machine in the dirty, uncontrolled and stressful environment of industrial life. This means that the laboratory device must not only be made and tested so as to perform reliably, but also that it must be designed so that it can be manufactured under normal industrial conditions by workmen using the jigs and tools of a standard factory and working to the tolerances of ordinary industrial practice. Second, the new technology has to overcome the natural mental inertia and conservatism of the consultants and designers who have to recommend their clients to buy it, and the unwillingness of the clients themselves to get involved in a risky new venture—something they have no experience of dealing with, and for which they have no guarantee of success. There is the well-known phenomenon of the 'N.I.H. factor'—N.I.H. standing for 'Not Invented Here'. Third, the new technology has to show a clear economic advantage, and not just a marginal advantage, over the existing technology, whatever that may be. It is not just a question of a new device's being cheaper to buy or to run than an existing device—it must be so much cheaper or better that manufacturers are convinced they can sell the new device before they will expend the capital investment on setting up production facilities. Very often this leads to a 'chicken and egg' situation in which no one will manufacture a new device on a large scale until it has been proved on a large scale that it will beat its competitors in the market-place. The new technology will then only force itself into existence if it can either do some job that the existing technology cannot do at all, or if it can provide some social benefit which will outweigh its failure to command immediate financial support. And while these struggles are going on, the manufacturers of the conventional technology, aware of the challenge being offered, will be refining and improving their product to whittle away the advantages offered by the new technology. All these are clearly illustrated in the case of superconducting technology.

Writing fifteen years after the discovery of the hard, high-field, high-current superconductors at Bell Laboratories, Dr R. G. Scurlock, of the University of Southampton, a leading British authority on superconductors, says 'This dramatic discovery transformed the pipe-dreams of physicists (but not electrical engineers yet) into reality almost overnight.' And referring specifically to the 'chicken and egg' problem, Professor Nicholas Kurti, in his introductory chapter to the standard work, *Cryogenic Fundamentals* (1971), writes: 'To break this vicious circle a new attitude of mind regarding cryogenics is called for. First of all, it must be realized that the wasteful, irreversible dissipation of energy (by heat and resistance losses in conventional electrical machines) is not ordained by any law of Nature, that it can be reduced and even made to vanish by various means and in particular by low temperatures. Nor is there any reason why the additional cost of producing and maintaining the low-temperature environment should not be more than compensated by the increased efficiency due to cooling.'

The long process of getting superconductivity out of the laboratory and into our daily lives began with what even staid scientific papers admit were some 'spectacular failures'. It had been shown that superconducting magnets could produce fields of many tens of thousands of gauss by the end of 1961. The new magnets were not just erratic, they were 'horrible things to work with', they had 'catastrophic breakdowns', according to John Hulme, still leading the Westinghouse research team. The main problems lay in the unreliability of the magnets and the very difficult materials, niobium-tin and niobium-zinc, with which the wires were made.

A superconducting magnet maintains its superconductivity by keeping out all external magnetic fields through setting up shielding supercurrents whenever a magnetic field tries to get in. But any change in the magnetic flux—as when an external field changes—causes a change in temperature of the superconductor. And the higher the temperature of a superconductor, the less it superconducts, and therefore the less it can provide shielding currents to keep out any outside magnetic field. Obviously, outside disturbances can develop into an avalanche situation and the whole superconduction system can break down catastrophically—the experts call this breakdown 'flux jumping'. It is also clear that unless this vicious circle can be broken at some stage, superconducting magnets will be inherently unstable.

The first signs of a solution came in 1965 from the Americans, Laverick and Stekly. They started embedding the superconducting wire in large quantities of pure copper. This meant that, if at any point the superconductivity broke down, the copper would carry the large current in a sort of bypass, allowing the superconductor to cool down again until it could again act superconductively. This was called 'cryostatic stabilization'—for the copper, like the superconductor, was necessarily at liquid helium temperatures at which its resistivity is even lower than usual. And with this technique the first reliable superconducting magnets were designed.

However, the disadvantage of the system is immediately obvious—the system is almost back to square one, because very large quantities of copper have to be used. Because the superconductor is so much better a conductor than copper, ten times as much copper as superconductor are needed to carry the current when flux jumps occur. So the magnet windings became heavy and large again.

The answer, rather surprisingly, was to make the superconductor wire extremely thin—a tiny filament only a few thousandths of a millimetre in diameter—and to have large numbers of these filaments embedded in a fine copper matrix. In this solution, the shielding currents set up by each superconducting filament are reversed in direction as an invading magnetic field passes across the filament. The average change in the magnetic flux level right across the whole wire is therefore very much less, and the heat generated is consequently reduced. Furthermore, the superconducting material has a much better contact with a far larger area of copper, and copper is extremely good at carrying away heat. The whole system is therefore much more stable, and complicated variations on this theme of tiny superconducting filaments in a matrix of copper are the modern solution to the problem of making reliable and trustworthy superconducting wires.

The remaining problem of such wires is that, since the copper is itself a very good conductor of electricity, it can 'couple' electric currents flowing superconductively in the filaments and behaving exactly as the designers hope. This coupling, or connecting, of the electric currents will theoretically upset the pattern of shielding currents in the individual filaments, in a very complicated way which would not be easy to sort out. A very simple solution emerged—and emerged by chance. It turned out that a simple twisting of the wires, by forcing the filaments into a helical pattern within the wire, completely cancelled out the internal coupling forces. Scientists believe they understand how this happens, but, on the whole, they are simply prepared to accept their luck and get on with using the simple solution. At the Rutherford Laboratory, near Harwell, in England, they found that wires that were behaving thoroughly erratically in their normal condition could be transformed into models of reliable superconductivity simply by giving them a few twists with the hand. And whereas some small magnets they were developing had been performing at only 10% or 20% of their designed capacity when using untwisted wires, they had only to make them with twisted wires to get them up to the full expected field.

Making a wire consisting of tiny filaments embedded in a matrix of copper is never very easy. When the filaments have to be of such brittle and intractable materials as niobium-tin the difficulties become almost insuperable. Throughout the 1960s metallurgists and engineers were searching for new ways to make superconducting wires. The first really satisfactory long lengths of wire were made from niobium-zirconium. The starting point for the manufacture was exceptionally pure niobium, melted by the application of an electron-beam. In the electron-beam furnace, iodide-zirconium is added in enough quantity to give 25%

zirconium in the final alloy. The ingots from the electron-beam furnace were then clad in jackets of stainless steel or molybdenum, and these ingots reduced to wire through four normal metallurgical processes—hot extrusion, hot swaging, cold swaging and drawing to wire. Finally the wire was heat-treated up to 700 degrees to improve current-carrying capacity by changing the internal structure.

But after a few years niobium-zirconium was replaced in favour by niobium-titanium alloys, containing just under half-proportions of titanium. The processes for producing this wire are similar to those just outlined, but the resultant alloy turned out to be more ductile, more easily manageable, easier to coat with copper, and above all able to remain superconductive in higher magnetic fields.

Niobium-tin retained some support, however, because of its high critical temperature, and quite exotic metallurgical techniques were used to produce it. In one process niobium chloride and tin chloride, heated until they turned into gases, were mixed with hydrogen and reacted together in a chamber through which a thin stainless steel strip kept at 1,000 degrees was passed continuously. The reaction deposited the mixture of niobium and tin on to the steel strip. Another process developed by General Electric in the USA was called the diffusion process. Here an initial strip of niobium is formed and a layer of tin of just the correct thickness is placed on it by dipping or evaporation or electroplating. The combined strip is then treated at 1,000 degrees to allow the tin and niobium to react to form niobium-tin.

Yet another group of methods involved essentially making the magnet by winding wires of niobium and tin and then treating the whole device so that the niobium and tin reacted together to form the superconducting alloy only after they had been put into magnet shape. The first such magnet ever made was started with a tube of pure niobium about half an inch in diameter. This was filled with powdered niobium and tin in exactly the right proportions to make the desired final alloy. The whole tube filled with powder was then drawn down to wire about one-tenth of an inch in diameter and formed into a coil. The whole coil was heated to form a successful superconducting magnet with wire-windings of correctly formed niobium-tin. A number of variations of this technique were developed, but all have now been discarded as obsolete, because, although there were a number of successes in the shape of superconducting magnets capable of producing 70,000 gauss and more, if there was any mistake in the formation of the winding or in the processing the whole thing became a total write-off.

Because niobium-tin is still the best superconductor where the requirements are really 'hard', work is still going ahead on improving methods of manufacture. Probably the most highly favoured nowadays is the production of niobium-tin ribbons in which layers of niobium-tin are formed by diffusion processes on a ribbon of niobium, with the whole coated in copper. This gives a fairly flexible material.

Even more advanced methods have been demonstrated, though they have not yet been put into commercial production. One of these involves vaporizing niobium and tin in the correct proportions with a plasma-torch and spraying the two materials together in layer after layer on to a copper cylinder which is at first masked with properly-wound copper wire. The wire is then stripped away to leave the niobium-tin correctly formed. A surface coating is then sprayed on to the alloy and the process repeated until a complete coil is formed. It has been shown recently that niobium-germanium is an even 'harder' superconductor, with a critical temperature as high as 23 degrees above absolute zero. At the moment it is considered impossible to manufacture this material into any usable shape, but it seems reasonable to predict that some mode of manufacture will eventually be found.

The commercial success of developing the first industrial process for large-scale manufacture of such composite and superconducting wire came from two British teams, one from Rutherford Laboratory under Martin Wilson and one from Imperial Metal Industries under Tony Barber. Their success was announced in 1969 and throughout the 1970s this type of material has been commercially available as the number of applications of superconductivity slowly but steadily expanded. Their first commercially-marketed superconducting wire contained sixty-one filaments of niobium-titanium superconductor in a copper matrix, with the filaments nicely and regularly twisted. The whole wire could be obtained insulated in plastic and a single wire was half-a-millimetre or less in diameter. Their process was promptly patented throughout the world.

In the years that have followed, this line of development has gone forward steadily until at the latest enquiry it was possible to fabricate kilometre lengths of filamentary composite wire with the superconductor filaments formed of that difficult material niobium-tin, and there are experimental lengths of this material containing no fewer than 40,000 niobium-tin twisted filaments in a 2-millimetre-diameter wire.

This extraordinary material is formed by taking a bronze ingot—bronze being an alloy of copper and tin—and drilling it along its length with 37 holes into which are inserted 37 niobium rods. The bronze ingot is then reduced to a rod by the normal metal-working extrusion or swaging processes and then drawn into thin wire. Then by a heating process the tin from the bronze is diffused into the niobium until niobium-tin is formed, leaving a highly coppery bronze as covering and matrix for the whole. The wire containing its 37 niobium-tin filaments is then cut into short lengths and put into holes drilled through another bronze ingot. This is again reduced in size and drawn into wire. A twist is given to the filaments at each stage, and the process can be repeated until the finally required result is reached. The limits to this at the moment lie in the heat treatments necessary to keep the bronze workable—these could eventually damage the superconducting capabilities of the niobium-tin alloy.

So the problems of making superconductive technology into a manufacturing possibility have been solved—there have had to be comparable developments in making the technology of liquid helium refrigeration a workaday routine job at the same time.

But at first only science laboratories and the British and American navies were serious customers for superconducting machines. Certainly, the first commercial sales of superconducting magnets were as parts of scientific apparatus. These superconducting magnets were small, and that was their attraction. They have been sold mostly as the essential core of devices called nuclear magnetic resonance machines. The nucleus of each atom can be regarded as a tiny magnet, and it will spin in a certain direction according to the magnetic field in which it finds itself. In so doing it will affect the behaviour of neighbouring atomic nuclei. By using their knowledge of such phenomena, scientists can analyse the structure of very complicated molecules and perform other important measurements. But obviously the magnetic forces exerted by the nuclei of atoms are very small and it needs a large magnetic field working on very small samples to produce effects which are large enough to observe and measure. At the same time, most laboratories are fairly small places and do not have room for very large pieces of equipment.

Taking all these factors together, the small superconducting magnet makes a very attractive proposition—it can produce a high field in a small space, it does not require large supporting power supplies or water cooling equipment, and many laboratories have the supplies of liquid helium necessary to bring the magnet down to superconducting temperatures. So it was possible to build comparatively simple machines that would fit into an ordinary scientific laboratory, with a single one-man control console, on which nuclear magnetic spin resonance measurements could be made. This is to take but one example of the type of scientific machine for which a superconducting magnet was ideal. The success of a small company such as Oxford Instruments, founded in a wooden shed in his garden by a member of the team at the Clarendon Laboratory, and specializing in making just such scientific measuring equipment, proves the point. And it was in this field that superconducting technology made its first commercial appearance.

The demands of the great accelerators were equally scientific, but very different. Nuclear accelerators are very large machines—a whole laboratory with a staff of several hundred scientists, engineers and technicians is built round just one or two such machines. In the machines streams of the particles that go to make up atoms—protons, neutrons and electrons—are accelerated to very nearly the speed of light and smashed into stationary targets or into each other to discover what they are made of. The acceleration is performed in enormous concrete tunnels buried underground; some are circular with diameters of several hundred yards or even of several kilometres in the case of the largest; others are huge straight pipes more than a mile long. The particles are kept on their tracks and focussed into thin beams by large numbers of very powerful, very carefully engineered magnets. The power consumed by such an accelerator is equivalent to the consumption of a small city. If the magnets could be reduced in size and run with virtually no power at all the savings in capital expenditure and running costs would be enormous. The opportunity of turning to superconducting magnets for such nuclear accelerators has been obvious to scientists for a long time, but the requirements for accelerator magnets include extreme reliability, for the machines are run for days at a stretch.

In fact, the first successful use of a large superconducting magnet came, not in running a large nuclear accelerator, but in the devices used at the output end of the machine. These are called 'bubble chambers' and are most easily thought of as giant cameras weighing many tons, centred round a tank containing gallons of liquid hydrogen or neon or similar material, which enable the sub-nuclear scientists to obtain photographs of the tiny particles as they crash into each other. In 1968 the Argonne National Laboratory in America started working the first bubble chamber with a superconducting magnet. This magnet produced a field of 18,000 gauss in a chamber nearly four metres across. It cost 2.4 million dollars, the same as a conventional magnet of the same size would have cost—but it saves the laboratory 400,000 dollars a year in cutting power costs. Following this same line of development, superconducting magnets were built for bubble chambers for the Brookhaven accelerator in the USA (it produced 30,000 gauss) and in 1970 for a bubble chamber at the European accelerator run by CERN in Geneva, which provided 70,000 gauss.

During the 1970s a new generation of even larger accelerators has been built, notably at Stanford and Fermilab in the USA at CERN, Geneva, and at Serpukhov in Russia. The possibility of using superconducting magnets in the main accelerator has been seriously discussed for many years, and certainly in the case of CERN detailed designs were drawn up for using superconduction. The new technology was not in the end selected for use, on grounds not only of questionable reliability but also, probably, because the capability to manufacture superconducting magnets on a fairly large scale was not available in industry. (Large-scale industrial production is a very different thing from producing one specially designed magnet, one-off, for a single bubble chamber.)

But now, at last, it seems that the big step forward has come. At Fermilab, not far from Chicago, the world's largest nuclear accelerator is having its power doubled—or possibly its energy costs and running expenses reduced. For on top of the present accelerator they have started installing the Energy Doubler/Saver. This consists in essence of an accelerator with large superconducting magnets, each weighing about eighty tons. They will be used to double the power of the accelerator, enabling the scientists to study even more exotic sub-nuclear particles (and perhaps to discover the elusive, but theoretically predicted, quarks as the basic building-blocks of all matter), or alternatively to halve the energy needed to achieve their present power, and thus cut their running costs and reduce their fuel consumption in a world where energy-gaps are beginning to become political realities.

But however much superconductivity may have forced itself into the world of the scientific laboratory as a commercially viable technology at the frontiers of knowledge, that is a far cry from producing and transporting electricity or power among ordinary citizens of the world. The world's first large-scale electric motor using superconductivity has been built and worked in an industrial environment, but the major interest in producing driving-power by superconductivity is confined to navies.

The superconducting motor is based on discoveries made by the greatest of all pioneers in electrical science, Michael Faraday. As far back as the 1830s he showed that if a disc of conducting metal was rotated with a magnetic field at right-angles to it, a voltage was generated between the centre and edge of the disc. Correspondingly, if a disc in a magnetic field has a voltage across from its centre to its edge, it will rotate. This is the principle of the superconducting motor, although at present the superconductivity is confined to the windings of wire that create the magnetic field. One important thing to realize is that such motors are d.c. (direct current) motors, and therefore entirely different to the electric motors of most of our modern electrical machines and gadgets which operate on a.c. (alternating current).

Before superconductivity became practicable there had never been a commercially successful motor of this d.c. type—it is called a homopolar motor—because such enormous quantities of heavy magnetic iron had been required. But there had been laboratory motors on this principle and the main features of performance of such a motor were known. In particular, such motors produced very high torque—that is, driving power—at very low speeds; this is not a common feature in motors of any kind, where usually maximum torque is produced at comparatively high revolutions.

The characteristic of producing maximum drive at low speeds of revolution makes such a motor particularly attractive as a mode of ship propulsion, where the propulsive efficiency of the propellors in driving the ship through the water and the increase of water resistance with speed follow laws which do not apply to land transport.

It was indeed at the behest of the British Royal Navy that the first attempts to build a superconducting motor were started by IRD which is based in Newcastle-upon-Tyne. This was in the early 1960s, before the stabilized filamentary superconductor wires could be made—so the decision had to be in favour of a direct current motor. Eventually a 3,250 horsepower motor was built to a full industrial scale and tried out in real industrial conditions at the Fawley electric power-station near Southampton. It was used to drive a standard pump and was mounted deep in the base of the power station in an area where water often collected round its foundations. It was not a complete success, but the troubles were more with the compressor than with the motor, and, when the world's first industrial superconducting motor was finally removed, it had proved its reliability and power enough to convince both the British and American navies to mount major programmes of research and development into powering their ships with the new technology.

The advantages of superconducting motors to drive naval vessels appeared to be fourfold. First, because the main shaft to the propeller no longer had to be connected directly to the diesel engines, entirely new layouts of ships became possible. Only the superconducting motor has to be mechanically connected to the propellers—the main electricity supply can be brought to the new motors by cables from anywhere in the ship. Secondly, all sorts of combinations of main engines, which are in the role of electricity generators, and the final drive of the superconducting motors could be envisaged, giving extra margins of safety against damage by enemy action and also giving economies in engine usage and fuel consumption. Even further fuel economies were the third advantage foreseen, because with electric-motor drive it is possible to use the engines in the most efficient manner for the particular power being demanded from them. Finally, superconducting motors should give much greater manoeuvrability than the gas-turbine engines of so many modern ships. Gas turbines cannot be reversed normally, so these ships need expensive reversing gears or reversing propellers.

The superconducting motor at Fawley power-station provided enough information for the Royal Navy to commission the design of an entire transmission system for a minesweeper hull. This was built and shore-tested in 1974, although design had only started in 1970. There was to be a complete system for one shaft of an operating minesweeper (the other propeller being driven by conventional means). There was a diesel engine and a superconducting motor for final drive. It is widely known that this system got as far as shore-testing, in a shape that would produce one megawatt of power. What is not so widely known is that the Royal Navy 'went the whole hog' and also included a superconducting generator to turn the diesel-engine output into electricity. However, at the shore-testing stage this system ran into many difficulties, as might be expected with any such revolutionary new design. The increasing economic crisis in Britain and the ensuing cuts in public expenditure forced the decision that it would not be justified to put the minesweeper superconducting power system into any actual ship. An important factor in reaching this decision was that while the hardware was being developed and tested, there had been parallel design studies on much more powerful systems. There had also, of course, been a great number of developments in superconductive technology outside the naval field going on all this time. It was quite clear that the bigger system could circumvent the problems that had beset the minesweeper prototype system, but the whole design concept had changed. So the one-megawatt system was never installed in a ship.

Meanwhile, the US Navy has mounted a major programme on superconducting ship propulsion systems. And as the British economic situation is still not likely to produce large funds for naval experiments, the Royal Navy is waiting to make a fresh assessment of superconducting propulsion which will be based largely on the American experience.

The basic physical law governing the efficiency of ships is that the resistance of water increases as the cube of the speed the ship is travelling through the water—on land and in air transport, air resistance increases only as the square of the speed of travel. The developers of superconducting naval propulsion units have been doing many studies of super-tankers, container-ships and the like. They foresee ships of the future with revolutionary propulsion systems. The three standard diesels of the super-tanker would each be attached to generators for producing electricity. This would be fed to superconducting direct current motors which might be rotating at as little as 60 or 40 revolutions per minute. Because the drive between diesel engine and propeller would not be directly mechanical, the main engines and the driving motors could be physically separated into different compartments, allowing changes in ship design, and diesels, generators and superconducting motors could be connected in a variety of alternative ways giving greater flexibility for maintenance and overhaul while running and much greater safety margins in case of accident.

The proponents of these superconducting motors also foresee their being used in other applications where high power and low speed are needed together—such as in driving the heavy hot and cold strip mills in the steel mill or driving pit-shaft wheels and lifts in deep coal, or gold, mines. The same sort of superconducting device run as a generator rather than a motor, could produce very large amounts of direct current electricity—perhaps as much as a hundred megawatts—and in this role would be very useful providing the power for large aluminium smelters.

But in most homes and factories the electricity we use is alternating current and as alternating current changes from one phase to another, the magnetic fields it sets up change in direction too, and there is continuous change in the quantity physicists call magnetic flux. It was precisely the change in magnetic flux that troubled superconductors in the early days of their development, and we have seen how the early superconducting magnets blew up when there were changes in magnetic flux. It was therefore considered unlikely that superconductivity would be applicable to normal heavy electrical engineering—at least, that was a widespread opinion in the early 1960s. This was one reason for the slow acceptance of superconductivity in industrial research and planning.

With the coming of the stabilized, filamentary superconducting wires, all that has changed. Many experts believe that the greatest of all the changes that the new technology will bring will be the superconducting power generator. And it is in this field that the

Americans have made their greatest contribution and done most of their work.

Much of the basic research has been done at the Massachusetts Institute of Technology (MIT), where the world's first superconducting a.c. generator was built under Professor Smith's leadership. It was a tiny machine, producing only one kilowatt of power—just enough to drive a single-bar electric fire. The MIT engineers have since built much bigger machines. But the leading contender in the field is almost certainly Westinghouse, where John Hulme's team has built a five-megawatt machine. This was completed in 1972 and has been steadily tested ever since, with all aspects of its performance being investigated. Broadly speaking, it has produced no surprises and it has shown that there are no inherent and insoluble problems in the superconducting a.c. generator. In this machine the heavy rotor of the conventional turbo-alternator is replaced by a d.c. superconducting magnet. The whole machine is very much smaller and lighter than the equivalent conventional generator.

The crucial question now is whether such a superconducting generator will prove economically worth while. The cryogenic apparatus, the refrigeration and the supplies of liquid helium are all expensive and their cost has to be balanced against the advantages obtained through superconductivity. The chief advantage of the superconducting generator is its power-density—that is, that it can produce more power from smaller size than the conventional machine. The present day turbo-alternator is limited in its power output essentially by the mechanical strength in rapid rotation of the magnetic iron that is its core and also by the amount of magnetism that iron can carry. The normal limit of operation of iron cores is about 20,000 gauss, and then the iron becomes saturated with magnetic force. But a superconducting magnet would provide 50,000 or 100,000 gauss and therefore enormously increase the power output. This in turn would cut down the relative capital cost of the machine. A superconducting power generator should also be more efficient as a machine, because the conventional alternators need 1% of the mechanical power available simply to make the rotor rotate, the equivalent of losing ten million watts of the potential electrical output of the largest alternators being designed nowadays; in a small superconducting set the chief losses for similar output would only be

200,000 watts spent on keeping the refrigeration of the liquid helium going.

These factors will be traded off by the designers of the next decade as they decide whether to build superconducting generators or to stick with development of the present sets. But there is one factor, not at first obvious, which will weigh in the scales to the advantage of the superconducting set. The latest turbo-alternators generating 600 or 1,000 megawatts are very large in size; the physical problems of handling them, such apparently mundane matters as actually getting them installed or maintaining them once they have been installed, are becoming increasingly difficult to master. A superconducting generator, despite its cryogenic complications, is estimated to be about one-third easier to build, handle and maintain, and this may prove conclusive in its favour. The best estimate is that prototype superconducting generators of virtually full commercial size will be built in the 1980s, and it seems likely that superconductivity will here have a big enough margin of advantage to force itself into everyday use.

But the future appears to be rather different for superconductive transmission of electricity. On the face of it, this seems the most obvious use of superconductivity, which literally offers transmission of electricity without any resistance. In Britain the idea was taken up with great enthusiasm in the 1960s, for at that time a new large supergrid was being planned and built to link all the electricity power stations into one huge system which would guarantee supplies to the whole of the country whatever local breakdowns there might be. (Such a national grid is attractive to a small crowded country; it is hardly as economically attractive across the wider spaces of America or Russia). But there was growing resistance to the march of the overhead cables on their huge pylons by groups of people we would now call environmentalists. Protests became particularly loud when the grid system had to march across some area of landscape beauty such as the Cotswolds. An answer for the future seemed to lie in buried superconductive cables, although the expense of refrigeration stretching for hundreds of miles was plainly going to be large. The Central Electricity Research Laboratories therefore took a leading part in developing superconducting transmission cables. They very soon produced a rigid cable which would carry large quantities of power under superconductive conditions. This was developed into a less rigid form, for any practicable transmission system would have to go round some bends and corners. And a superconducting buried highpower transmission system was built which went a mile or two round the outside edge of the laboratory's site near Leatherhead.

But this alternative power transmission system did not sweep into the market. The economics of overhead cabling are very attractive up to loads of 1,000 megawatts—for the air provides virtually free insulation and heat sink facilities. Only the demands of environmental preservation argue against the system. And if our care for natural beauty forces us to lay cables underground, at least for short stretches of their run, the cable industry has been hard at work producing new and better alternatives to their present system. For the next generation of heavy-duty cables we shall probably turn to the 'gas-pipe' system—in which the electrical conductors run down the centre of a pipe filled with sulphur hexafluoride gas which acts as both insulator and heat-remover. Even in the more distant future it seems likely that aluminium or copper, cooled to about 80 degrees above absolute zero—roughly the temperature of liquid nitrogen—will provide good enough conductivity—or low enough resistance, the same thing—to show a more economical answer than transmission in liquid helium along superconducting cables at four degrees above zero.

So it has taken superconductivity between sixty and seventy years to move from being one of the most extraordinary discoveries ever made in the laboratory to being a practicable engineering, working technology. It is still only making a big impact in the market for the very 'highest' technology in the most esoteric areas of scientific research. There is little doubt in the minds of those best qualified to judge that, within the next ten years, it will have major impact in the heavy electrical engineering market and in the world of ship propulsion. Indeed, an American professor of physics has gone so far as to describe superconductivity as 'the greatest advance since the wheel'.

8 Cold at Work

When we think about heat energy and high temperatures and how we use them, it is natural to think first of supplying power to engines—raising steam, making electricity, driving cars and lorries and aeroplanes. We also use high temperatures for other purposes, notably to soften or even melt hard substances such as metals so that we can shape them and form them as we require. In another way we use high temperatures for cooking food and sometimes for baking things—bricks or bread—so that they harden. The coming of cryogenics in the second half of the twentieth century has meant that we can perform a great many different but equally useful functions by using cold and low temperatures. Cryogenic engineering—cold at work—uses low temperatures to render materials more brittle, or to turn materials into solids so that we can shape them into things we want.

Cryogenic engineering has sprung almost entirely from the industrial gas industry. The best way of separating the components of our atmosphere is by the use of cold, as we have seen. The end products, the separated gases, have emerged from the process at a very low temperature, and that part of cryogenic engineering which consists of putting the peculiar qualities of low temperatures to work has essentially capitalized on the coldness of the gases rather than on the gases themselves. Liquid nitrogen, oxygen, hydrogen and helium are used as carriers of cold in these operations.

The most spectacular of these uses of cold is freezing the ground for civil engineering work. Surprisingly, the first recorded use of this technique was in South Wales more than a century ago. There they wished to sink a shaft for a coalmine through a layer of waterlogged soil, so they sunk steel pipes into the ground in a ring round the spot. The pipes had steel noses, pointed so that they could be driven right through the waterlogged layer. Then cold brine was pumped down all the pipes so as to form a volume of frozen ground round each pipe. The process was continued until the cylinders of frozen earth round each pipe joined up to make a continuous and solid ring holding water and wet earth out from the centre. The shaft was then begun within this ring-fence of frozen columns of mud by digging through the still soft and unfrozen earth at the centre.

The principles used at the South Wales coalmine have remained in operation since then, though the techniques of freezing—the cryogenic engineering—have been vastly improved. The use of cold brine is still common, though nowadays it would probably be calcium chloride brine which is still liquid down to about minus 27 degrees. And now there would be an electrically-driven transportable refrigeration unit feeding cold brine to a storage tank. From this tank the brine can be delivered by motorized pumps to all the steel probes or pipes from a delivery pipe that runs right round the circle. Not a great deal is known about the strengths of various soil types when frozen, but it is agreed that the crushing strength of soil rises rapidly as the temperature is lowered and the strength at minus 15 degrees is probably about twice that of soil only just frozen. The economic considerations governing the engineering decisions are then to balance the soil strength needed against the extra cost of achieving lower temperatures.

The alternative method of freezing soil—and one that is especially useful on small cramped sites—is to pump liquid nitrogen down the tubes or probes and let it freeze the soil as it evaporates. This technique was used when the Post Office had to perform a comparatively small excavation that was very close to a corner of St Paul's Cathedral, where it was important not to disturb the foundations of that famous building.

A more normal use of the technique is typified by an operation in Newcastle-upon-Tyne in 1974. Here a main sewer shaft was being dug when a liquid layer of sand and water was encountered at a depth of six metres below the surface. Progress was halted completely until the soil could be stabilized by freezing. Thirty-five steel tubes, each twenty-seven feet long, were sunk all round the mouth of the shaft and then liquid nitrogen was pumped down the tubes. Two tanker-loads of liquid nitrogen were used each day for three days, until the ground was completely frozen. Then it had to be kept frozen for a week while the excavation went through the middle and retaining walls were built to hold out the water. The whole operation consumed four hundred tons of liquid nitrogen. One particular problem was that all inspections at the bottom of the shaft had to be done by men wearing breathing garb and air bottles, because the atmosphere down the shaft consisted almost entirely of nitrogen evaporating through the freezing soil.

Theoretically it should be possible to perform a soil-freezing operation by connecting all the probes together underground and running a complete refrigeration system through them. There would be obvious problems in making the system leakproof and the whole thing would be very expensive—but in particular the Russians apparently built and used a plant of this sort in the 1960s.

The largest-scale use of ground-freezing techniques has been in the construction of new tanks to hold liquefied natural gas—usually using the LNG as the coolant. These experiments have, however, run into a number of problems.

An interesting variation of the soil-freezing theme is being used currently as part of an investigation designed to help preserve salmon fisheries in Welsh rivers. The Welsh Water Development Authority was worried about the effects of big land drainage works on the gravel and sediments in river beds where salmon spawn. But almost nothing is known about the sort of conditions that salmon like for spawning or the factors that affect the successful hatching of the eggs. The only thing to do was to examine sections of river bed where it was known that salmon actually chose to spawn.

Three rivers were chosen, the Gwili, the Dulais and the Cennen, all in the county of Dyfedd. To take samples of the river beds at spawning places, steel stand-pipes were driven into the chosen sections of bed to a depth of about forty-five centimetres. Then liquid nitrogen was poured into the stand-pipes until the whole section of the river bed was frozen solid. When the stand-pipe was hoisted out by a light tackle, the section of river bed came with it and was immediately taken off to the laboratories for study. The sections removed were about ten feet long and two feet deep. The holes in the river bed were filled with stones and gravel, but they will be reexamined this year, using the same technique of recovery by freezing, to see what changes occur in the natural course of river flow. One factor that is being most carefully watched is the level of silt on the river bed, for it is known that salmon avoid areas of deep silt, because the silt can suffocate the spawn.

In a rather similar manner the cold of liquid nitrogen has been used to cope with an unusual fire emergency which occurred in Derbyshire, when a hopper containing seventy tons of pulverized coal—the fuel for a kiln—was found to be smouldering and smoking. Immediate action was taken to stop the fire spreading, by covering the top of the coal with tons of stonedust and by dowsing the whole hopper with water. This contained the fire but the temperature of the coal remained at 300 degrees—and that is high enough to start coal combustion again.

The final answer was to stick a twenty-foot-long hollow steel lance into the coal and pump liquid nitrogen down it. The low temperature of the gas brought the temperature of the coal down to a safe level within twenty-four hours, and since nitrogen is inert there was no fire hazard.

A rapidly growing family of uses of cryogenic techniques exploits the fact that low temperatures harden and embrittle many materials. The fastest growing of all industrial applications of cryogenics lies in this field. There is rubber and plastic de-flashing, for instance. This simply means getting rid of the unwanted 'flashes' of plastic or rubber which invariably remain as very thin flakes of the material clinging precariously to the edges of objects which have been moulded or pressed into shape. The flashes are the surplus material that has been squeezed out between the edges of the dies or moulds. Getting rid of them by hand or by old-fashioned trimming methods tends to be labour-intensive and costly. The cryogenic method of de-flashing uses the low temperature of liquid nitrogen to render the thin sheets of flash brittle and frozen, while the main rubber or plastic part is cooled but not frozen. The parts are then simply tumbled in a motor driven barrel; the flashes break cleanly off leaving a smoothly finished object. A similar approach is providing a new method of dealing with some of the most difficult forms of scrap and waste, and even enables much material to be recycled for profitable use. One of the most refractory forms of waste in our motorized world is the tyre—huge piles of worn-out tyres can be seen on the edges of most large cities. Sometimes they are burned, and the air is polluted by plumes of thick, black, greasy smoke. The rubber itself cannot be saved and recycled in most cases, but it can be embrittled by cold, and so after an application of liquid nitrogen the tyres can be smashed by hammers and literally reduced to grains and dust. Many people will have seen the classical demonstration in which the lecturer dips a tennis ball in liquid oxygen and drops it on the floor where it smashes into pieces; this illustrates clearly the principle involved in dealing with old tyres.

A more constructive approach to dealing with scrap, using exactly the same cryogenic principle, has been developed by Belgium's largest scrap merchants, George et Cie. The process takes in low-grade metal scrap, such as wrecked cars or old washing-machines, which are compressed into bales and then cooled with liquid nitrogen. These embrittled bales are fed into a giant 'fragmentizer' which reduces them to bits of metal about the size of potato crisps. Furthermore, at this stage all the contaminants such as copper, zinc, aluminium and non-metal parts are shaken free from the mild-steel which is the basic material of such objects as cars and washing-machines. The steel fragments are small enough to flow freely down pipes or chutes, and can be remelted and turned again into a high-quality steel.

Many manufacturing processes require materials to be ground into small, identically-sized powders or pellets—plastic materials are often required in this form before being fed into moulding machines or extrusion presses. In many cases, manufacturers would like to have their raw materials in this form, but the materials are held to be ungrindable, because of the dangers of fire or explosion or because the materials would react with oxygen in the air during the grinding process. And all grinding processes are limited by the amount of heat produced, so that attempts to speed up the flow of material through the machines all too often end with the material melting and clogging up the whole system. Cryogenic grinding, in which the material to be ground and the grinding machinery are cooled by a liquefied gas, is claimed to provide the answers to most of these problems. The gas companies usually suggest using liquid nitrogen because the gas evolved is inert and will both protect the particle from reactions with the atmosphere and reduce flame and explosion risks. Most of the big gas companies are pushing their cryo-grinding systems. In a rather similar vein they offer systems for temporarily hardening—though not embrittling—rubber so that it can be worked by fairly conventional cutting tools.

Everyone associates heat with expansion—but it is equally true that cold causes contraction. This is the principle of shrink-fitting, which is one of the new and valuable techniques of cryo-engineering. The application is obvious—if two parts of a machine have to fit together closely—such as an axle fitting into the centre of a wheel so that the two will rotate together—then either one can heat the wheel so that it expands, pop the axle into the centre hub and let the wheel cool to fit back tightly on to the axle, or one can cool the axle so that it contracts, fit it into the wheel and let the axle expand back at normal temperature until it fits tightly into the wheel. Cryogenic engineers claim that shrink-fitting is much superior to the heating process for two chief reasons: first, shrink-fitting is cleaner and less dangerous than heat treatments; secondly, and more important, cooling produces no distortion in manufactured parts whereas it is virtually impossible to guarantee that heat will be applied uniformly and that any piece of work will expand absolutely uniformly.

Again, it is the industrial gas companies that are pushing the shrink-fitting technique in manufacturing industry, though often it is not the gas itself they are selling, but the cold carried by the gas. The claims made by all of them are similar, for instance Air Liquide, which calls the process *'l'emmanchement par contraction'*, claims to have shrink-fitted a four-ton steel cylinder on to the outside of a tank, while British Oxygen, for its part, claims to have had success with shrink-fitting a huge shaft into a trunnion. For more repetitive jobs they point to a gearbox assembly process where a manufacturer found he was having a 65% failure rate in fixing a seal between two parts; the failure rate was reduced to nil when shrink-fitting was used.

The part played by cryogenics in space technology has mainly been concerned with the production of liquid oxygen and liquid hydrogen as rocket fuels. But there is one small area where the cold rather than the gas has mattered, and that is in the construction of environmental test-chambers. Space conditions are themselves cryogenic in that temperatures are very low once our earth's atmosphere has been left behind. It becomes essential, therefore, to test all materials and parts and entire satellites in the conditions of extreme vacuum and very low temperatures that they will meet. These conditions can be created in very large test-chambers in which space machinery can be validated, and obviously very considerable quantities of liquid nitrogen, liquid hydrogen and even liquid helium are used to keep temperatures down. Furthermore, in achieving the vacuum conditions in these chambers the new techniques of cryopumping are being developed, although it is quite clear that the end of this line has not yet been reached.

It has long been an industrial practice to have 'cold-traps' in systems designed to produce high-vacuum conditions. These cold-traps are surfaces, cooled by liquid nitrogen usually, on which vapours in the system will condense because of the low temperature, and thus be drawn out of the evacuated area. To achieve even smaller pressures, the same idea is used on a larger scale and at a lower temperature. Then almost any gas left inside the system after the pumps have finished will condense on to the cold surface and can be taken away. In many cases cryopumping has proved to be the only way to achieve very low pressures in the large volumes needed for space simulation.

A much less obvious use for cryogenics is to get rid of 'noise'. Noise in this case is the great enemy of the electronics engineer—it does not mean noise that the human can hear but the word is used in a sense derived from our normal hearing. Noise is normally defined as unwanted sound, to distinguish it from the organized and patterned sound that we regard as speech communication or music. Of course, music and speech can sometimes be unwanted and we invariably describe the transistor radio producing pop music in the peace of the countryside as 'noisy'. To the electronics engineer noise is an unwanted signal, invariably random and unpatterned, produced by the natural vibration of the atoms and molecules in the atmosphere and in the material of his apparatus. Since these vibrations—which are what we also describe as heat energy—are reduced as the greater orderliness of low temperatures is attained, so electronic noise is reduced by low temperatures.

From this, it follows that whenever an electronics engineer is trying to receive very faint or very weak signals, he will use a device which is cooled to low temperatures. The low temperature reduces the noise made by the materials in the receiver and this increases the 'signal-to-noise ratio'. The two usual devices or receivers used in these circumstances are the mavar or the maser. The first of these—the mavar—is a parametric amplifier, the most sophisticated version of the normal radio amplifier which receives the weak signals and increases their strength before passing them on to the main receiving apparatus. The second works on the same principle as the more dramatic and better-known laser, using the principle of stimulated re-emission. Both devices are normally cooled with liquid helium. At the centre of most of the large dish aerials which receive and send telecommunications to communications satellites—the dishes at Goonhilly in Cornwall are good examples—there is a metal tripod with what looks like a metal box fixed at the top. This metal box is, in fact, positioned at the focus of the reflecting surface of the dish, and it contains one of the supersensitive receivers. Invariably these receivers are cooled by liquid helium, so cryogenic engineering is playing its part in most of the telephone calls we make across the Atlantic or to even more distant continents such as Australia.

A similar application of cryogenics affects mostly the military field. It is not quite accurate to classify infra-red detectors as electronic instruments, but they are very similar in many ways. They usually depend on crystals of esoteric materials like indium-antimonide to detect infra-red radiation, and these will only work successfully when cooled to at least the temperature of liquid nitrogen. The military importance of infra-red detection depends on the fact that engines, being hot, produce infra-red radiation, and this radiation can be used to 'attract' missiles fired from other aircraft or from the ground. It is also possible to make infra-red-detecting fuses for shells, so designed that the shells—normally anti-aircraft shells—will explode only when close to an engine or other heat-producer. A missile can be made to home in on an aircraft engine by using its infra-red detector, so this must be kept in operation for several minutes, needing a continuous cooling apparatus. But an infra-red fuse need only operate for a few seconds, so they can be supplied with smaller cryogenic devices which give a brief 'one-shot' cooling effect. Another military use of infra-red detectors, and therefore of cryogenics, is surveillance by night. This can be a simple matter of providing an individual soldier with an infra-red 'night-sight', but it can also involve fitting a surveillance satellite with infra-red detectors to cover vast areas at night. For such a long operation as the latter, bottles of liquid gas would be far too heavy for use in satellite or spy-plane, so a minute liquefier would probably be carried. A Stirling-cycle refrigerator, such as that developed by the Dutch firm of Philips, will probably provide the long-term answer to this problem—but here we approach the borders of national security and little has been published about the solutions to these technical difficulties.

Certain types of laser, such as those using gallium-arsenide, will emit beams of coherent infra-red radiation when cooled to cryogenic temperatures. These have military applications for guiding bombs or missiles and for communications purposes and range-finding. Masers, which also need cryogenic cooling, as noted above, are used in missile-detection systems. And all these military applications impose special, and often difficult, cryogenic requirements. The cooldown time for an infra-red fuse, for instance, may be required to be as little as two seconds, while the duration required of satellite infra-red sensors may be as long as three months. When the solutions to these problems emerge from under the cloak of security, they will surely find even further applications in industry.

Far removed from the semi-secret glamour of space-technology and missile development is another application of cryogenics which has already been applied in a commercial way—indeed, it is claimed that the first truly commercial industrial use of a superconducting magnet has been in this field of filtration. Filtration is an essential part of such mundane businesses as sewage treatment, water purification and the preparation of raw materials. And the important new advance of magnetic filtration ideally requires cryogenic help in the shape of superconducting magnets.

The principle of magnetic filtration is quite simple—all the fluid is driven through a layer of stainless steel wool, which is continuously magnetized up to its magnetic saturation point by a powerful magnet, almost inevitably a superconducting magnet. Nearly all mineral particles—ordinary dirt—will be magnetically susceptible enough to be attracted on to the magnetized steel wool. Even molecules of substances that go into solution in water—things like nitrates and phosphates—will be pulled out by the magnetism, and so will the heavy metals such as mercury that present an ever-increasing pollution threat to our modern world.

To catch particles that are non-magnetic, especially living organisms such as bacteria, it is necessary to seed the liquid with colloidal iron or small grains of mineral such as magnetite. These will either be ingested by the bacteria, rendering the whole organism susceptible to the magnetic attraction, or cause the particles or bodies to coagulate with them in the condition known as colloidal suspension, and then the whole mass will be pulled on to the steel wool. The great advantage of this filtration system is claimed to be the high flow-rates that can be achieved through the steel wool magnetic filters. And this high flow-rate becomes more important as the big cities use more and more water and produce more and more sewage—Boston alone has to handle more than 300 million gallons of sewage a day and Manchester, in England, uses 130 million gallons of purified water each day.

A variation of this technique is to use magnetic steel wool filters to attract particles of metal ore, when the objective is to collect metal, rather than to remove it from water. In many cases where low-grade ore is being mined, the rock is crushed into small pieces mechanically and then carried away for further processing by water. The ore in water makes what is called a slurry, and if this slurry is passed through a magnetic filter most of the valuable metal-bearing particles are collected and concentrated. By using a similar procedure English China Clays, in Cornwall, are purifying the clay they produce. There is some commercial secrecy surrounding this process, but it is known that it uses a superconducting magnet—the first to be bought commercially.

One final application of cryogenic engineering must be mentioned. It produces devices which may appear to be only of technical importance to scientists, but which are of enormous basic value to all industry, using some of the very latest discoveries in superconducting science.

The essence of this latest development is a device called the Josephson junction. This is a phenomenon discovered by a young research student in Cambridge even before he had obtained his doctorate. And this discovery earned Brian Josephson the Nobel Prize for physics in 1973. A Josephson junction is, in essence, a small piece of insulating material connecting two superconductors. When alternating current flows in both superconductors the electrical behaviour on one side of the insulator is affected by the performance of the parameters of the current on the other side. It is said that the superconducting electrons 'tunnel' through, or across, the insulator—though this is only to be taken as a figure of scientific speech.

Josephson's work was at first primarily a calculation, a prediction of the effect, published in 1962. Probably the most important features are that below certain levels of current the voltage across such a junction is zero and, correspondingly, if a voltage is placed across such a junction a current will flow across it. But there are now many forms of Josephson junction, depending on the thickness of the insulator; some even have ordinarily conductive connections between the two superconductors.

The whole subject has developed far beyond Josephson himself, but many of the developments were predicted in his first great paper. 'It is probably true to say that there have never been more predictions crowded in the space of two pages,' writes L. Solymar (1972). There are nowadays Josephson tunnelling, Josephson junction, d.c. Josephson effect, Josephson radiation, Josephson plasma resonance. There are other important phenomena produced by Josephson junctions, such as quantum interference effects, but all are dependent on the cryogenics needed to produce the superconductivity.

Josephson's discovery is interesting in itself as an example of the scientific process, and how that process does not follow the rules which scientists believe govern their own work. In fact, the Josephson effect had been noticed thirty years before, in 1932, by Holm and Meissner. But there was no theory then to account for what they had seen, and the results seemed to be entirely incompatible with classical physics (as indeed they were). So no further follow-up was made and no one advanced any explanation for the effects. The whole thing was forgotten in the press of other more comprehensible work.

In 1960, however, the time was ripe. A body of theory had been built up following the production of the BCS explanation of superconductivity. Josephson's first work was entirely prediction from this theory, though he rapidly showed by experiment that some of his predictions were indeed true.

The importance of Josephson's work to date is that it allows extremely precise measurements to be made of the very units by which we normally express the quantities of electrical phenomena. Using the junction devices it is possible to measure voltages as low as 10^{-15} volts and magnetic fields as low as 7×10^{-11} gauss. Extremely accurate instruments have been made such as the SLUG—the Superconducting Low Inductance Galvanometer. Above all, Josephson effects can be used to examine superconductivity itself. Solymar (1972) writes, 'The most important application of superconductive tunnelling is diagnostics. We use superconductive tunnelling to reveal the properties of the superconductive state. It is true to say that the spectacular development of superconductivity in the last decade owes a lot to the simple technique of tunnelling; in fact superconductive tunnelling is the most sensitive probe of the superconductive state.'

But Josephson junctions and tunnelling are opening the way to the even more exciting development of superconducting electronics—normally referred to as SCE. SCE are likely to provide the first and largest impact to these ideas on our normal everyday lives. It may be a matter of regret, but it will come as no great surprise, that it is in the sphere of weaponry, or euphemistically 'defence', that the development of the new technology has started making great strides.

The Josephson junction in the form of a 'squid ring' is the most sensitive magnetometer ever invented. (A magnetometer is any device for detecting or measuring changes in a magnetic field; 'SQUID' is an acronym for Superconducting Quantum Interference Device.) It is predicted that a squid ring magnetometer should theoretically be able to detect a nuclear submarine under water from a satellite orbiting above the earth. The US Navy is working hard to perfect a system of several such superconducting magnetometers which could be mounted in satellites or aircraft, or even on a land-base, and which would enable them to detect and track the movements of submarines. A nuclear submarine is large enough, and metallic enough, to appear to the squid ring magnetometer as a bar-magnet moving against the background of the earth's magnetic field. It would create, in the jargon of the defence-electronics world, 'a detectable magnetic anomaly', which means that, even allowing for the minor changes and variations and fluctuations of the earth's natural magnetic field, it ought to be possible to detect the extra changes caused by the presence and movement of a missile-carrying submarine.

At the moment the balance of deterrent between the two nuclear super-powers is held by the missile-submarines, which, because they usually cannot be detected, are the invulnerable platforms from which either America or Russia can strike the other after the comparatively vulnerable ground-based missile sites or comparatively slow nuclear bombers have been eliminated.

It is not unreasonable to suspect that emergence into prominence of the Cruise missiles, and the continuance of SALT talks centring on the new missiles, may be connected with the realization that missile-carrying nuclear submarines could soon lose their virtual invulnerability. This would also affect the internal balance of power between the navies and air forces of the nuclear nations.

A large number of other military devices representing enormous advances in the techniques of warfare can also be predicted, because superconducting electronics offer lower losses, higher sensitivities, greater speeed and wider bandwidths than almost any components of conventional electronic technology.

SCE components could have big advantages over present-day components, particularly in the world of highly sophisticated radar-tracking and automatic following of aircraft, missiles and ships, simply because they promise to be smaller, to work faster and to need less current. But it is also believed that they could be used in the very difficult business of communicating with submarines under water. Here they offer an attractive alternative to such schemes as the American Sanguine-Seafarer system which would require many hundreds of square miles of territory to set up underground 'aerials' for sending communications through the earth and the waters of the oceans to their submarines. Using much smaller transmitting aerials and accepting their much weaker signals, but providing the submarine with a set of highly sensitive squid ring magnetometers which could be towed behind in a 'fish', offers a realistic solution to the problem. The concept has been the subject of tests which proved that signals could be received several hundred feet below the surface of the sea.

In electronic counter-measures (ECM—the systems used in enabling planes, missiles or ships to avoid detection by enemy radar), the superconducting devices are believed to have an important future because of their extraordinary sensitivity and speed of reaction. In simple terms, ECM systems mounted in, say, an aeroplane, detect the beam of the enemy radar as it passes across the plane and then send out a signal of their own to jam the radar beam or give a false reading. The defending radar operators counter this by continually changing the frequency of their signals or by building a frequency code into their beams so that they know which are their own signals bouncing back from an aircraft and which are false signals. The ECM devices in the aircraft try to follow the radar changes in frequency as fast as they can, so that they can duplicate the changes or codes in the signals and falsify the returns with their own output. This is an interminable game, but obviously if SCE components can react much faster than conventional components to changes in frequency, then they have an enormous future in this highly classified, but vitally important area.

Superconducting electronic devices also offer advantages over normal electronic components for a wide range of less exotic operations. A device called an SCSO, a superconducting cavity-stabilized oscillator, has already been invented. It is claimed that it is unmatched by any other device in several important respects for performing the function of an electronic 'clock'. A clock, or oscillator, which will keep perfect time for long periods is an essential part of many electronic systems—even a simple radio system needs its basic clock for keeping the correct time at whatever frequency it is operating, so that the system can detect those variations in the basic signal which actually carry the message and which can be retranslated, by a receiver with a matching clock, into the words or music we want to hear. And apart from regulating the internal timekeeping of most electronic systems, an accurate electronic clock is needed for the sophisticated world-wide navigation systems based on satellites. Satellite navigation systems, of course, give nuclear submarines their accurate positions for launching missiles, but they are also coming more and more into use for civilian operations such as fixing the precise location of merchant ships, or of oil-drilling platforms far out at sea.

Similarly, it is foreseen that SCE devices could take the place of many of the present day 'low noise amplifiers'—the masers and mavars which are essential features of the military and civilian communications systems that work through communications satellites. These satellites are normally placed 22,000 miles above the earth, where they orbit at such a speed that they stay above one point on the earth's surface and appear, to us below, to be stationary. As we have seen on p. 204, supersensitive amplifiers are therefore needed which will concentrate on improving the weak signal we want, and which will not only reject 'noise', unwanted signals, but also will not generate too much 'noise' in their workings. Here is another field where SCE may provide better answers than our present systems.

The biggest problem facing the SCE devices lies not in reaching standards of performance that outmatch present devices—it is achieving their potential in a reliable way and carrying their own cryogenics, for they all work only at liquid helium temperatures. Achieving reliable working performance is a problem all new devices have to overcome. It is known, at least in principle, how to manage this difficulty and how to overcome the fairly standard problems. A great deal depends upon improving manufacturing standards and achieving quality control over mass output. Carrying the cryogenic necessities, in both the economic and engineering senses, is the outstanding obstacle to SCE development. But there is great confidence that refrigeration cycles can be made small and reliable, and there are many new developments known to be available. There is also at least one old development, the Stirling engine, which has long been in need of a technological niche in which it can find fulfilment, and this niche will probably be offered by the cryogenic demands of future SCE systems. There has also been much rather exotic cryogenic engineering in military contexts which can be turned to the use of SCE systems. For instance, in the 'one-shot' cooling devices used in infra-red detectors, a liquefied gas is released to cool the vital infra-red-sensitive crystal down to its working temperature for just long enough for the missile to do its work. Although liquid helium temperatures are not required for infra-red detectors, the work that has gone into such systems could surely be adapted to SCE needs. The infra-red detection systems of orbiting military satellites have already been adapted to civilian use in satellites which take infra-red pictures of crops and other features of ground interest. Again, it seems reasonable to expect that this space cryo-engineering could be developed to provide the low temperatures needed for super-conductive devices.

Perhaps the best guarantee that all these hopes for SCE devices are not just wishful thinking is provided by IBM, a corporation that is not renowned for wasting its efforts on over-fanciful research projects. In 1976 a scientific paper was published by a group of researchers at IBM's Thomas J. Watson research centre in Yorktown Heights, New York, which revealed that the computer giant, which is also one of the world's largest electronics manufacturers in its own right, was developing the techniques which are known to be necessary for the production of many SCE components. The essential features are achieving small sizes and accuracies greater than any so far in production use. IBM is achieving these standards by using an electron beam to cut or etch out the necessary circuits and contacts. And in their final summing-up Broers, Molzen, Cuomo and Wittels (1976) claim that their technique 'should allow the direct fabrication of many useful devices', among which they specify 'superconducting or semi-conductor tunnelling structures'. A Josephson junction is a superconducting tunnelling structure.

Moving away from the shadow of nuclear warfare, it can be pointed out that virtually all the applications for SCE in the military field have civilian counterparts—radar, communications, electronic clocks are all used in the world outside defence, and it is normal for components and devices which are first used in, and developed for, the military field eventually to percolate into the civilian field where we all feel the immediate benefit. Such will surely be the case with superconducting electronics.

There is, however, one clearly civilian use of SCE which is already in the first stages of development. There is much use in medicine of the electrocardiogram (ECG) to examine the performance of human hearts and of the electroencephalogram (EEG) which can detect the electrical activity in the human brain. Both can be life-saving devices in the diagnosis of serious illnesses. But both have the disadvantage of requiring the patients to be connected up directly to the recording machines. This means attaching electrodes on to or just into either the skin of the chest and back or in the case of an EEG, forehead and neck—a sticky and uncomfortable, if not actually a painful process. By using squid ring magnetometer devices, however, it will be possible to record the electromagnetic signals from heart or brain without ever touching the patient with wires or probes. This system has already been shown to work satisfactorily at the Massachusetts Institute of Technology, and they have added an extension by measuring the performance of the patient's lungs as well. This is achieved by getting the magnetometer to follow the behaviour of a fine dust of magnetic particles which the patient inhales with his air supply.

It is possible, in the view of the MIT research team, that there could be even more surprising developments in knowledge by further application of squid magnetometers to the study of the human body, for virtually no work has ever been done on the magnetic fields of the human body, or on the reactions of the body to small magnetic fields.

Finally, there is an even more startling prediction made about the use of SCE. Because these devices can be made very small, because they require little power, because they produce little heat, and because they can work very fast, it will be possible to make extremely small but very powerful computers working at liquid helium temperatures. Indeed, the logic units which are the essence of all computers can be made so small, if we use superconducting electronics, that any one of them can occupy less space than a human brain cell. It should be possible to assemble as many SCE logic units as there are cells in the human brain, in the same sized space as is occupied by a human brain. And the computer that is so formed will have units that work a million times faster than any cell in a human brain. The thinking machine may be in sight. It will need refrigeration such as no brain needs, but it could start taking shape within five years.

The technology of supercold may well be able to examine the human body without touching or probing it, and may even imitate the human brain, but doctors do not have to use such extremes of low temperature for treating the ailments of the human body. Temperatures much nearer to zero Centigrade than to zero absolute will satisfy the requirements of medicine. However, medicine also uses liquid nitrogen, and the temperature of minus 196 degrees that it provides. Certainly, in most medical applications of cold and low temperature, the cooling action, is provided by liquid gases, which are used as carriers of low temperature, or as cooling agents, rather than in their own right as oxygen or nitrogen or carbon dioxide. But the use of cold in the service of healing began long before the industrial gas manufacturers began supplying liquid oxygen or nitrogen to hospitals and doctors.

In the fourth century BC the Greek, Hippocrates, knew of and wrote about the use of ice and snow to stop haemorrhage and to reduce pain and swelling. The ice bag continues to be used today in the corner of the boxing ring. The classical tradition was continued in AD 1,000 by Avicenna, who also mentioned cold as a stupefacient. And in the late Middle Ages an English 'leech book' advises patients to sit in cold water before the leeches are applied around the pock marks. But probably the true starting point of modern medical interest in the application of cold to curing diseases came with the publication of *De Nivis, or The Medical Use of Snow* by Thomas Bartholin in 1661.

Bartholin's most important pronouncement was that his master—Severino of Naples—had discovered the application of snow caused numbness and that he used this cold-caused insensibility to help in many incisions and in preparation of such operations as cauterizing ulcers. This seems quite reasonable to the modern scientific mind, certainly a wiser approach than that taken at the start of the nineteenth century by James Currie, who doused his patients with buckets of cold water from the River Mersey to reduce their fever. However, Currie did note later that bathing with tepid water had more effect on cooling a fever, and this sort of lukewarm wash is still given often to feverish children.

A chief surgeon to Napoleon's armies, Dominique-Jean Larrey, seems to have been much more observant about the true effects of cold on the human body. He is credited with the discovery that refrigeration can produce anaesthesia, largely on account of his noting that his assistants kept dropping the instruments in the appallingly cold weather during the battle of Eylau in February 1807. Larrey in fact does not seem to have pursued this idea, and after the disastrous Russian campaign, when he had plenty of opportunity to learn about cold, he wrote, 'Cold exercises its sedative effects principally on the brain and nervous system'—a remark which can be accepted today as perfectly accurate.

By a strange quirk it was the coming of chloroform that brought cold seriously into the mainline of medical thinking. Those who opposed the new technique just had to be able to offer the patient some alternative to the new 'painless' surgery—something more than just pointing out that fifty deaths had recently occurred with the use of chloroform in operations. James Arnott was one of the leaders of the anti-chloroform school. He offered cold instead, and campaigned throughout the 1850s in a series of lectures and pamphlets and medical papers which now occupy a place in medical libraries. In 1852 he wrote on 'benumbing and congealing temperature for inflammatory, painful or malignant disease.' Two years later it had become 'on benumbing cold as a preventive of pain and inflammation from surgical operations, with minute directions for its use'. By 1858 there are other serious medical papers offering 'insensibility to pain in dental operations . . . painless surgery by congelation'.

Arnott made important advances in his field, using crushed ice, snow and salt in a bladder or rubber bag applied locally to the skin to produce numbness or analgesia. The method was taken up in the USA as well, notably by J. Mason Warren in New England. Sprays of ether or other rapidly evaporating liquids were also applied to the skin to provide immediate rapid cooling locally, and it was found that total anaesthesia can also be produced by cold. Another practitioner hit on the very sensible idea of getting his patients to suck ice cubes to numb their throats before he started investigations with his laryngoscope. The use of ice and snow attracted military surgeons and medical men both in America and in the outposts of the then British Empire. And there has been a continuous, though minor use of cold ever since. This rather pragmatic approach culminated in 1938 when Temple Fay in America started reducing patients' whole body temperature to between 80 and 90°F in cases involving severe gangrene or great pain and shock after severe injury. Several others also reported success using this technique of hypothermia in extremely bad wounds or accident cases.

It was only after the Second World War, however, that there came to be much scientific basis for the 'cold' approach. The events of the war and particularly studies of the effects of intense cold on shipwrecked seamen started a line of studies and experiments on the real physiological effects of cold on the human body. These studies went on into the 1950s and at that time were proceeding parallel to the studies of Andjus and Audrey Smith on whole-body freezing of animals. The two lines of study together started to influence medical thinking and to provide both reasons for the use of cold and some knowledge of the limits of cold the body would resist.

The new knowledge of the physiology of the body under cold conditions, the lower rate of metabolism of the cells, the reduced demand for oxygen in brain and blood, led to the first dramatic use of hypothermia. This was in cardiac surgery, the 'open-heart' operations. In these procedures the body was deliberately cooled, often to temperatures as low as 10–15°C. The patient was first anaesthetized to cut out all the normal body reactions to cold, such as intense shivering. Then the body was immersed in cold water and in some cases cooling liquids were introduced into the body cavities. Finally the bloodstream was led out of the body and into a blood cooler through which it flowed steadily. Most of of the preliminary experimental work was done by W. G. Bigelow on dogs until the procedures were proven enough to move on to human beings. (His main publication on the subject appeared in 1958.) Soon surgeons in many centres were able to stop the heart and brain for as long as eight minutes while they performed delicate operations on the heart.

Towards the end of the 1960s came another major step forward—this time it was cryosurgery. Probably the best-publicized of these operations is cryosurgery of the brain, started by I. S. Cooper in the USA as early as 1962. The object of these operations is still chiefly to cure the effects of Parkinson's disease, the tremors and rigidities that afflict mainly elderly patients. The method was by the destruction of small groups of brain cells, known as basal ganglia, by the direct application of cryogenic liquids or gases to the cells. In principle the operation is very simple—a fine tube is placed precisely on to the target cells and liquid nitrogen is dripped down the tube. The liquid nitrogen usually mostly evaporates at the bottom end of the tube but the intensely cold vapour which comes out is sufficient to destroy the basal ganglia. Obviously the details of such a delicate brain operation, particularly the precise positioning of the tube, or hollow needle, are fairly complicated and sophisticated, but the operation has been performed very widely. Cooper himself published a study of 1,000 patients treated in this way, showing that the rigidity and tremors of Parkinsonism had been overcome in more than 90% of cases and that there had been a mortality of only 1%. There is some medical controversy over the long-term advantages of the operation, and there are alternative chemical or radioactive ways of destroying the cells, but the operation still remains one of the prime examples of cryosurgery.

The fact that extreme cold congeals cells has been known for many years, as we have seen. The use of cryogenic 'knives' to make 'bloodless' incisions has become widespread practice. But the use of this congealing power has had most effect on ophthalmic surgery, where cryogenic techniques have been widely used for reaffixing detached corneas. The rival to cold here is the extreme heat of the laser, which can be used for the same purpose.

Cryosurgery is also used on the eye for the clean removal of the lens in cataract operations, and for treating certain advanced forms of glaucoma.

The destructive effects of extreme cold on living cells is now being applied to the removal of many malignant tumours, and has even been used for hysterectomy. Many medical equipment companies manufacture special instruments or whole ranges of instruments for cryosurgeons. These are often in the extremely convenient forms of 'pencils' or 'pistol-grip' devices, because the object of the equipment is usually simply to put a thin tube into contact with the tissue to be destroyed, so that the cryogenic liquid can be poured or squirted down the tube on to the cells. Thus it is often possible to perform a cryogenic operation without making incisions into the skin, simply by inserting the cryoprobe into one of the orifices of the body and introducing the cryoliquid when the probe touches the organ, for example a prostatectomy can be performed in this way.

These cryosurgical operations all sound rather fierce and technical. But the techniques can be used equally successfully for such comparatively simple and mundane operations as the removal of tonsils or the destruction of piles (haemorrhoids).

Cryosurgery, like cryogenic engineering, is very much an expanding and exciting subject. A quick look at the programme of any modern congress or symposium on cryobiology or cryosurgery reveals that new types of operations and medical procedure, based on either the extreme cold of truly cryogenic liquids or the less extreme hypothermic techniques, are regularly being announced. The one goal that still eludes the cryobiologists in general is the successful storage of whole organs so that patients who need transplants can be immediately supplied from an organ-bank. But a great deal of scientific work is going on to achieve this aim, and those working in this field are reasonably hopeful that they will succeed. If they do, the storage will require major cryogenic installations.

9 Towards the Limits of Cold

After he had liquefied helium and discovered superconductivity, Kammerlingh Onnes continued struggling to reach even lower temperatures. He installed larger and larger pumps above his cryostat and pumped away the vapour as his liquid helium boiled off. But he never reached a temperature much below one degree above absolute zero. The story of how later scientists got further down the temperature scale is a story of science, pure and simple. It is not only the story of reaching even-lower temperatures but also the story of another extraordinary phenomenon which has been found near absolute zero—superfluidity.

There is no question of useful applications or engineering at these temperatures. Nor has anyone found any practical value in superfluidity—yet.

It is worth remembering two things before approaching the region where absolute zero appears to be in sight. First, at this level of coldness we must forget our everyday ideas about temperature, where we think of every degree of temperature as being an equal step up or down. Below one degree, scientists cheerfully talk about thousandths or even millionths of a degree above absolute zero. The lay reader must realize that this simply means that a substance is a hundred times or a thousand times hotter than another substance (as was explained in Chapter 7). Secondly, it is important to remember the Third Law of Thermodynamics. In its simplest form, Nernst stated that science can never reach absolute zero and that as a substance gets nearer and nearer to absolute zero this means that it gets nearer and nearer to a state of perfect orderliness—the mysterious quantity of entropy, disorder, becomes less and less. Telling the 'story' or 'history' of the approach to the unattainable absolute zero necessarily imposes a logic upon events. It may be a narrative logic or a chronological logic that is imposed. And this logical progression is the way the scientist likes to see it—A showed that such and such occurred and three years later B was able to measure the precise temperature of the occurrence; from this C deduced that such and such was happening and published this theory in 19xx—reference 3; this was confirmed by D in a great series of experiments extending over ten years, and so on. Any author's narrative tends to confirm this view of what happened. But the logic is an unreal one; a logic imposed by hindsight. Science is not the steady piling-up of logical progressions. And the true story of cryogenics shows this. It is a tale of muddle and confusion, and of results which not even their own authors dared to admit, because they contradicted all known theories. We have seen that cryogenics produced such impossibilities as electricity circulating without any resistance, discovered because the thermometers were giving apparently ludicrous readings. We have seen that the Josephson effects were spotted forty years before Josephson predicted them but were ignored because there was no theory to account for them.

Similarly, Kammerlingh Onnes almost certainly saw superfluidity ruining his attempts to reach lower temperatures—but he did not know it; who could blame him for not expecting a liquid to run upwards inside the walls of his containers?

The great Dutch scientist described his final attempt to lower the temperature of helium (by pumping off the vapour as the liquid boiled) in 1922 to the Faraday Society in a communication titled, *On the Lowest Temperature Yet Obtained*. He concluded that unless another substance more volatile than helium could be discovered, the process had reached an end. And yet he rejected that same logical conclusion and declared a programme to prove himself wrong: 'the first thing needed is long and patient investigation of the properties of matter at the lowest temperatures we can reach.' He set himself along this road. He made discoveries he could not understand and hardly dared to publish. But the route he pointed out has been followed. Further consideration of the basic properties of matter did provide a way to still lower temperatures. These lower temperatures showed new and unsuspected basic properties of matter. These properties allowed still lower temperatures to be reached. We do not know whether the end of this road has been reached, but the story of the penetration into new depths of cold is the story of the interplay between these two studies. And the present-day laboratory production of extremely low temperatures—temperatures of the order of a few thousandths of a degree above absolute zero—is performed in the interests of studying properties of matter. Yet scientists can and do produce temperatures of only a few millionths of a degree above absolute zero by using the properties of matter discovered at the 'higher' temperatures of thousandths of a degree above zero.

Onnes died in 1926 in the month of February. Two months later, on 9 April, the American Chemical Society heard from the Californian, Professor Latimer, that one of his young research students had an idea for reaching temperatures below those of Onnes's liquid helium. In December this young man, the Canadian William F. Giauque, published his ideas in detail. Quite independently, Peter Debye had sent an identical proposal to the German *Annalen der Physik* in October of the same year. The situation was very similar to the rivalry between Cailletet and Pictet to produce the first liquid oxygen fifty years before.

The cooling system proposed by Giauque and Debye is called magnetic cooling and it depends essentially on the concept of Nernet's Third Law of Thermodynamics, which states that, as absolute zero is approached, entropy tends to zero as well. So the object of magnetic cooling is to produce a greater orderliness in a specimen and thus achieve lower temperatures.

Any atom consists of electrons orbiting around a nucleus which consists of protons and neutrons. The electrons not only orbit around the nucleus but each one has its own spin, just as the earth has its spin while orbiting the sun. At low temperatures with so much of the normal movement reduced, the spin of the electron becomes more important. These can be regarded as making each electron 'look' like a little bar magnet. It is possible to align the spins mostly in the same direction by applying a strong magnetic field to a specimen—this aligns all the little bar magnets in the same direction and is the process we recognize as magnetization of a substance. In theory, then, according to the Third Law, at absolute zero all the electron spins of a substance would be lined up in the same direction in perfect orderliness, or complete absence of disorder. And certainly it is known that the opposite is true—that heat reduces magnetization. This is explained by saying that the atomic vibrations—heat energy—disorder the electron spins and make every electron spin in a different direction. In the early days of the twentieth century Pierre Curie had done a great deal of work on this matter and had shown that the magnetic susceptibility of many metals and alloys, that is, their liability to be magnetized, decreased as the absolute temperature increased, exactly as we should expect if heat energy, appearing as atomic vibrations, disorders electron spins. (Although Pierre Curie is most widely known for his work with Madame Curie on radioactivity, his work on magnetism is equally important and many materials have their 'Curie point', the temperature at which magnetism is destroyed in a specimen of the substance.)

With many substances the application of Curie's law about magnetic susceptibility decreasing with increase of temperature does not work perfectly because their electrons tend to pair off, in couples with opposite spins, each pair behaving like twin small magnets sticking together, north pole to south pole. It is, therefore, more difficult for an external magnetic field to align all the spins in the same direction. In a group of comparatively rare substances, salts of the rare earths and iron, the crystals contain a number of electrons which cannot be paired with an opposite-spin counterpart, and these electrons are isolated from each other so that they have little effect on other unpaired electrons. These substances are particularly interesting for scientists studying the application of Curie's law, and it is a matter of some irony that Onnes himself studied one of them, gadolinium sulphate, very carefully in the last years of his life. With his colleague, Weltjar, he cooled gadolinium sulphate to about one degree above absolute zero by immersing it in liquid helium at the lowest temperature they could reach. They found that gadolinium sulphate still obeyed Curie's law down to one degree—i.e. its magnetic susceptibility continued to increase as they lowered the temperature. The irony lies in the fact that they failed to appreciate the applicability of this observation to achieving even lower temperatures, though it was this point that Giauque and Debye seized upon. For if gadolinium sulphate is still obeying Curie's law at one degree above absolute zero, there must still be some unpaired electrons which can be aligned by an external magnetic field at that low temperature. This amount of disorder or entropy still remaining at a temperature where the atomic vibrations of heat energy are very small gives the scientist something to work upon.

Both Giauque and Debye proposed to impose order on these random electron spins when the temperature of a crystal of gadolinium sulphate had been reduced as far as liquid helium could take it. Then, if the atomic vibrations were so small that they could not affect the ordered electron spins, the entropy after magnetization would be lower and the temperature would be reduced. And this is exactly what happened—though it took seven years to do it.

Seven years may sound a long time to achieve a comparatively simple operation. In fact the achievement, which was largely Giauque's since Debye was a theoretician, represented the extreme of experimental advance in its time. The liquefaction of helium was still, in the pre-war, pre-Collins days, a complicated business requiring expensive cryostats. And the essence of the operation was to subject a crystal of salt, inside the middle of the liquid helium cryostat, to an intense magnetic field. It was no easy matter to provide the intensity of magnetization required and it was still more difficult to concentrate the field of force at exactly the right spot since the magnets had to be outside the cryogenic apparatus. Finally there had to be a heat-switch, in itself a new invention.

The procedure planned by Giauque required, first, the liquefaction of helium and then the pumping off of helium vapour to reduce the temperature even below the 4·2 degrees at which helium becomes liquid. The salt crystals—gadolinium sulphate—were placed in contact with liquid helium which was at a temperature of about one degree above zero as pumps took away the vapour. The salt was then subjected to the magnetic field and the liquid helium was used for the first time as a coolant to carry away the heat of magnetization from the salt. This was achieved by placing the salt in a container which was filled with helium gas, the whole being submerged in a bath of liquid helium. So what actually happens is that the whole thing remains at a temperature of about one degree, but rather more helium boils off, carrying the heat of magnetization away with it. Helium gas is used around the salt because it is such a good conductor of heat. Then when all the electron spins of the salt are considered lined up in the same direction so as to achieve maximum order, the helium gas is evacuated from the innermost container. The salt is then in a vacuum and the connection with the liquid helium has been 'switched off'—this is the heat-switch. At the same time the magnetic field is taken away. The magnetic cooling is then realized in the salt, and it can be shown to be much colder than the liquid helium in the cryostat outside.

Giauque carried out his experiments on three days in March and April 1933, at the University of California. On the last, and most successful day—9 April—he achieved a temperature of 0·25 degrees (just a quarter of one degree above absolute zero). What is more, he held that temperature, or somewhere near it, for several hours.

Thus was opened the era of magnetic cooling, an era which continues up to the present. But Giauque had not only opened this new era, he had also set new technical standards in cryogenics, particularly in methods of insulation. When other laboratories, such as Leiden, followed in his footsteps they met many initial disappointments, notably in failures to hold the low temperatures they reached. Even in the years after the Second World War many magnetic cooling experiments were failing because heat seeped back into the salt at a rate far faster than could be accounted for. Scientific detective work showed that this heat flow seemed to be worse in the daytime than at night. Then it was noted that the failures seemed to occur more often if there were mechanical pumps working on or near the experiment. This provided the vital clue. It was shown that the vibration of the nylon threads holding the salt crystal in position were heating up the salt, or allowing heat to leak in. Naturally this vibration was worse during the day when all the machinery and traffic outside and inside the laboratories were in action. So solid supports of very low conductivity material had to be used to carry the salt.

The logical conclusion from Giauque's successful demonstration of cooling by imposing order on the magnetic directions of the electron spins was to impose even further order by aligning also the even smaller and weaker spins of the nuclei of atoms. This is nuclear magnetic cooling.

Nuclear magnetic cooling is not only the logical continuation of the arguments which led Giauque forward to his success. It is also achieved in broadly the same manner—but naturally, being the same only more so, the experimental complications are even greater technically. It was finally achieved by Simon and his colleagues at the Clarendon Laboratory at Oxford in 1956, almost twenty years after it had first been openly discussed by Simon and by Gorter, of Leiden, in the 1930s.

Simon had built up an extremely strong team in Oxford, based on men like Kurti and Mendelssohn who had worked with him at Breslau but had been forced to flee from Hitler's Germany. Simon himself, as a decorated officer of the First World War, was exempt from the anti-Jewish laws, but had been attracted to Oxford through his contacts with Lindemann, who became famous as Churchill's wartime scientific adviser, Lord Cherwell. So Franz Eugen Simon, holder of the Iron Cross (First Class), ended his life as Sir Francis Simon, CBE, who had been one of the leaders of the British atomic bomb effort known under the code-name of 'Tube Alloys'. Another facet of this extraordinary and much-loved man was that in his Breslau days he would probably have been the best equipped to put into practice the theories advanced by Debye and Giauque. Indeed, some colleagues thought he had been left behind in the race when Giauque achieved his success. In fact, Simon would have disapproved of anyone, including himself, who had raced Giauque to put Giauque's ideas into practice; it was against his view of scientific ethics. When the time came to put his own development of those ideas into practice in the shape of nuclear magnetic cooling, the war intervened. But when the war was over he returned to the field and, as we have seen, in 1956 he finally succeeded.

The temperature reached by Simon and his team was eventually shown to be 0·000016 of a degree above absolute zero—that is, they had reached to a few hundred thousandths of a degree above zero. But their results took a great deal of interpretation. It is nowadays agreed that they had indeed reached this temperature but only so far as the nuclei of the atoms are concerned. The remainder of the substance—technically called the lattice—is but weakly connected with the nuclei of the atoms and did not reach such a low temperature. This state of affairs continues today in similar very low temperature experiments.

Simon and his team had set off on their journey to the very lowest temperatures from a starting point of a temperature just a few hundredths of a degree above zero. Since Giauque's original opening-up of magnetic cooling, new, and more efficient, paramagnetic salts had been discovered which reached these lower temperatures when demagnetized. The Oxford scientists embedded the ends of some very fine copper wire in such a salt and reduced its temperature as far as they could. They then carried out their nuclear magnetic cooling on the free ends of the copper wires. At these wire ends they achieved their new penetration into even colder areas.

Very little further progress down the temperature scale has been achieved in the following twenty years. It is now possible to reach temperatures measured in millionths of a degree above absolute zero—but this only applies to the nuclei of atoms, usually of copper. The bulk of the material still remains at temperatures measured in milli-Kelvin, or thousandths of a degree above zero. However, there has been progress in reducing the temperature of bulk materials using entirely different methods of cooling (see p. 234).

The first uses of magnetic cooling after Giauque opened the field in 1933 were to experiment on and measure the behaviour of the para-magnetic salts themselves. In other words, Onnes's programme was being carried out—research on the fundamental nature of matter at very low temperatures. Gradually the work was extended to small samples of other materials that could be cooled by contact with the very cold salts. Most of the work concerned the specific heats of substances—that is to say, the way in which different sorts of matter absorbed or carried heat-energy at these very low temperatures. Other experiments also showed that many elements became superconducting at temperatures lower than one degree above zero.

The really exciting work came, however, with the study of liquid helium itself at the new low temperatures. In other words, the spotlight turned back on to helium which had come on to this particular stage only as an instrument, the only instrument, which could be used to lower the temperature enough to enable magnetic cooling to start being useful.

And once again the paragraph has to begin for the umpteenth time, with the words: this sprang from work started by Kammerlingh Onnes. That great pioneer had finished his scientific life with a series of experiments, carried out in collaboration with the young American, L. I. Dana, on the measurement of the specific heat of helium at very low temperatures, well below the point at which the helium became liquid. The results were not published because they seemed ridiculous—it was easier to explain them as a breakdown in the equipment rather than try and find a scientific explanation for them.

Onnes had in fact been investigating the properties of liquid helium at intervals ever since he had successfully established the fact that the gas could be liquefied at all in 1908. In that very famous experiment he had even made an estimate of the density of the liquid and found it much lower than he had expected. Two years later he returned to the subject and made further measurements, which showed that the density of his new liquid steadily increased until it reached the temperature of 2·2 degrees above absolute zero, when it started to decrease. He published these figures early in 1911, but he could not explain this phenomenon and he was distracted from the subject by the discovery of superconductivity. It is even considered possible that he had seen and observed an extraordinary new property of liquid helium but had been so baffled by it that he had not recorded it. In his Nobel lecture in 1913 he referred to his density measurements and suggested that various strange phenomena, as well as superconductivity, might be explained by the quantum theory of energy expounded by Max Planck in the first years of the century. We have seen the difficulty Onnes faced when dealing with such an 'impossible' thing as superconductivity. He was faced with other equally 'impossible' phenomena when he examined the physical properties of liquid helium. Kurt Mendelssohn, one of the great Oxford scientists of the generation that followed up Onnes's work, explains 'The scientist, who regards the happenings which he observes as evidence of an organic fabric, will always try to link up what he finds with the body of established knowledge. This makes it difficult for him to comprehend at first sight a new phenomenon which has no connection with known fact . . . Onnes had the satisfaction of knowing that in superconductivity he had discovered an entirely new aspect of matter which until then had lain hidden in the world of very low temperatures that he had opened up. Of another, equally strange, phenomenon he had a few glimpses without however grasping its fundamental significance. It was much the same story as with superconductivity; the very magnitude of the impending discovery prevented his making it.'

The new phenomenon was 'superfluidity'. It is a condition that has only been observed in liquid helium. It took scientists virtually the whole of the next generation—Mendelssohn's generation—to find it and establish it in the body of scientific knowledge. It is only now, fifty years after Onnes had failed to grasp the enormity of what he was seeing, that the third generation of cryogenic scientists are really delving into its mysteries. And even now the major manifestation of superfluidity is virtually inexplicable according to our latest and most advanced theories.

Kammerlingh Onnes well knew that there were further important discoveries to be made. He returned to the problem of the density of helium in 1924. He repeated his earlier measurements more carefully and even further down the temperature scale. There could be no doubt that liquid helium contracted—became more dense—as its temperature was lowered below its boiling point, minus 269°C, and that is exactly what we observe with other liquids at more ordinary temperatures. But he confirmed that below the temperature of 2·2 degrees the behaviour of helium reversed—it began to expand and its density fell. For his final great scientific effort Onnes followed this up with the help of Leo Dana. They studied several other physical quantities such as the latent heat of evaporation and the specific heat of liquid helium as the temperatures were lowered. When they published their work in 1926 the measurements were given only down to the temperature of 2·5 degrees above absolute zero—showing a steady and unspectacular decrease in the specific heat. We know they just did not believe the sudden dramatic change in their graphs around the 2·2 degree mark—when the reading varied wildly and increased beyond all belief.

It was six years more before Onnes's successor at the Leiden laboratories, William Keesom, published his own figures, which showed that it was not Onnes's equipment which had broken under the strain, but his self-confidence and imagination. Keesom showed clearly that at 2·2 degrees, the precise temperature at which the density of the liquid stops increasing and starts to decrease, the specific heat of liquid helium changes abruptly to a much higher value. The change was dramatic in the extreme—there was no graph or continuity between the values above 2·2 degrees and those below that temperature. We can see now that this meant there was a change in the state of the matter (liquid helium) being investigated; we can see now that there is a direct parallel with the equally sudden appearance of super-conductivity.

Keesom guessed correctly that there was such a change in the state of liquid helium at this critical temperature of 2·2 degrees. What he could not guess was the nature of the new state. Since no one had ever made helium into a solid, and all other substances could be shown to have a solid state, he assumed that the new state must be a version of solidity—but was faced by the problem that the helium was still plainly liquid below 2·2 degrees. The concept of a 'liquid crystal' was invoked, the idea being that the helium had in fact turned into crystal, or solid, state but that the crystals were so loosely connected that they slithered over and around each other and so appeared as a liquid.

Later investigations, consisting essentially of passing X-rays through liquid helium at temperatures below 2·2 degrees, showed that there was no crystalline formation in the liquid. Keesom could therefore only say that there were two forms of liquid helium—Liquid Helium I and Liquid Helium II.

Investigating these two states of liquid helium in the beginning of the 1930s, a team at Toronto (McLennan, Smith and Wilhelm) was trying to observe the boiling of liquid helium at very low temperatures. To most people it will sound a nonsense to talk of reducing the temperature of a substance and then watching how it boils—it is worth remembering that boiling, the transformation of a substance from the liquid to the gaseous state, is achieved not only by increasing the temperature, but by varying the combination of temperature and pressure; liquid helium, however low its temperature, can be made to boil by reducing the pressure above it. The actual boiling of liquid helium appears like the boiling of any other liquid; the liquid is in a state of violent motion and bubbles rise continuously to the surface where the gas is released to the container above and the amount of liquid is steadily reduced.

The Canadian team saw this all happening as usual until the temperature was reduced to the figure of 2·2 degrees. And then all the violent motion and the bubbling stopped—'the liquid became very quiet', they reported. This observation was simply included in the report on their work; they drew no conclusion from it. It seems certain that Kammerlingh Onnes himself must have seen this change and failed to understand its meaning. We can now see there must be a simple interpretation—the heat conductivity of liquid helium at this temperature suddenly increases a million times over, in a very similar way to that in which the electrical conductivity of a very cold substance suddenly increases by a colossal amount as superconductivity sets in. But it took the scientists of the 1930s a long time to realize this.

Keesom, in a sense, actually recorded the fact. Working with his daughter Anna, he repeated with even greater accuracy his measurements of the specific heat of liquid helium. They saw that, as the dramatic change at 2·2 degrees occurred, they were also measuring a change in the heat conductivity—but no one connected this with the change in behaviour seen in Toronto. Instead, Keesom started work to measure the physical characteristics of Helium II by itself. He constructed a version of the standard physical apparatus for measuring the heat conduction of a poor conductor. The experiments failed completely. And suddenly, because the fact could no longer be avoided, it was realized that liquid helium below 2·2 degrees must conduct heat far better than copper or other metals. There was no explanation possible on the theories of the time—indeed, there are only doubtful explanations even now. But science had to face the fact that at this very low temperature helium behaved like matter in an entirely new state, never before considered possible. It was very like the moment when superconductivity suddenly forced itself into the framework of scientific knowledge—but this was not superconductivity itself; it was something different and extra, for liquid helium is not superconducting, it is if anything an insulator.

The two or three years following this realization that liquid helium could show a major new phenomenon—the years from 1936 to 1939—saw a flood of new discoveries in the field. Strangely enough, nearly all of them were published in just one volume of the British scientific journal *Nature*—volume 141—shortly before the war clouds finally closed on Europe and all purely scientific work ceased.

As one looks back on those extraordinary results of the late 1930s, results which have never made any impact on the general public because they were overshadowed by and under-reported against the rising menace of war and the later dramatic events in nuclear science and electronics of the radar type, it is difficult to know whether the emotions of the scientists involved were sheer excitement or ever-increasing bafflement as Liquid Helium II proceeded to show that it would behave in exactly the opposite way from what is expected of a normal liquid. On several occasions teams of scientists, as Onnes had done, were led to doubt their own results, only to find that later work proved they had been right all along.

First, a group at the Cavendish Laboratory at Cambridge found that, in opposition to all previous results, heat would flow better through Helium II if the difference in temperature between the two ends of the flow was reduced—in all normal circumstances heat conduction is greater when the difference in temperature between the two ends of the flow is greater. This we all know intuitively—if you put something very hot next to something very cold the heat will flow 'faster' than if there is not much difference between the two temperatures. But if Helium II, or superfluid helium as we now call it, is the conductor, this is not so at all. So, shortly after Allen, Peierls and Uddin reported this from Cambridge, they themselves publicly cast doubt on their own results and it was not until ten years later that it was shown they had been correct with their original observation.

The Cambridge men were, in a sense, distracted by an even more extraordinary result which emerged from their own attempts to clarify their puzzling initial observations. They found that, unlike any other liquid known to science, Helium II actually flows towards any source of heat applied to it. The experiment had been conducted by putting a glass bulb with an open neck into a pool of Helium II, so that it looked like an upturned flower vase standing in a pool of water. When the bulb was slightly heated, Helium II flowed into it so that the level of liquid inside the glass bulb was higher than the level in the pool—and we all know this is exactly opposite to what we observe in normal conditions. Then the scientists almost blocked up the neck of the vase—but the narrower they made the channel through which the Helium II had to flow towards the heat in the bulb, the faster it came and the higher the level inside the bowl stood above that of the pool outside. Finally they broke open the 'bottom' of the vase (now standing at the highest point because the vase is upside down) and found a most dramatic jet effect—Helium II flowed from the pool, upwards towards the heated bulb and then shot right out of the open 'top'. They called this 'the fountain phenomenon'.

Properly, however, the fountain phenomenon should be called a 'thermo-mechanical effect'. It is a mechanical motion forced on to a mass of material (the Liquid Helium II) by heat. And shortly after the discovery of this effect, a group at Oxford showed also that there was a true reverse of this phenomenon. It is called the 'mechano-caloric effect' and it consists of a change of heat, or temperature, in a mass of matter, brought about by a mechanical force. More specifically, the Oxford scientists showed with a similar apparatus, consisting of an inverted bulb with its neck in a pool of liquid helium, that if Helium II was allowed to flow from a higher level inside the bulb through a narrow channel down to the pool of liquid below, the temperature of the helium would be reduced as it came out through the neck of the bulb. The two effects together showed that in Helium II the flow of mass and heat in opposite directions is connected in a unique and reversible way.

It was in Cambridge, at this same time, that the basic phenomenon of superfluidity was finally established and demonstrated as the most extraordinary of all the properties of helium below the temperature of 2·2 degrees. Both Keesom in Leiden and the Toronto group had been worried by leaks in their apparatus. Pieces of equipment that seemed perfectly tight to all normal tests proved useless through leakiness when the temperature was lowered below 2·2 degrees. Both laboratories demonstrated that the viscosity—the stickiness or ability to flow easily—of liquid helium appeared to decrease at these very low temperatures. Their results were based on the measurement of the decrease of the motion of cylinders or discs which were set spinning or swinging while submerged in Helium II. Unfortunately, the results published by the two laboratories differed somewhat in details.

But early in 1938 the Cambridge team of Allen and Misener showed that if Liquid Helium II were forced to flow in very narrow tubes (called capillaries) or in a very narrow gap between two plates there was virtually no viscosity at all—no resistance to flow. And the narrower the space in which the flow took place the less was the viscosity—so that it could be reduced to one-millionth of the normal value. The Cambridge results were published at the same moment and in the same journal—once again *Nature*—as identical results obtained quite independently by Kapitza in Moscow. (P. Kapitza had himself worked in the Cavendish Laboratory at Cambridge in Rutherford's nuclear science team before he had begun to concentrate on cryogenics after his forced return to Moscow.) It was Kapitza who termed the new effect 'superfluidity', but it was obvious to everyone working in the field that this must be connected in some way with the extraordinary and parallel increase in heat conductivity in liquid helium which occurred at exactly the same temperature.

The number of scientists working in this field, and the number of laboratories in which they worked, was still extremely small—there were not more than half-a-dozen places with the necessary equipment to perform cryogenic work. Whenever and wherever the results were published, all the centres kept in touch with each other by telephone, letter or personal visit. The string of revelations about the extraordinary behaviour of Helium II was circulated with great rapidity within this charmed circle—which provided an apt illustration of the concept of the 'invisible university' at work. So it was in the same year, 1938, that Oxford—in fact Mendelssohn, working with a research student, J. G. Daunt—showed the strangest effect of superfluidity. Helium II, half-filling a small glass beaker held over a pool of the same liquid, will actually climb up the sides of the beaker, move over the top lip, slip down the outside of the beaker and drip steadily into the pool of liquid below. To say that nothing like this had ever been seen before is to understate the case—it is a complete denial of everything that we all associate with the normal behaviour of matter.

The behaviour of Helium II under these conditions showed Mendelssohn what superfluidity really meant—it was not just that the liquid would flow better, the narrower the orifice through which it moved; it is rather the case that the liquid will flow along any surface with absolutely no friction whatsoever if it is allowed so to do. The Oxford scientists showed that the film of helium climbing out of their beaker was less than a hundred atoms thick—which is far, far thinner than any of the tiny tubes or gaps between plates of the Cambridge or Moscow experiments. And the speed of flow was absolutely steady—it was in fact a critical velocity which set in as soon as the temperature was reduced below 2·2 degrees. With experiments using a variety of beakers, one inside the other, containing various different levels of liquid helium, Mendelssohn showed that there was no eventual difference in level between the liquid in the beakers, and therefore no difference in the potential energies of the various masses of liquid. This is exactly equivalent to the absence of electrical potential in wires which are carrying superconductive currents.

When all these astounding results about superfluid helium were published just before the War, the theorists were able to get some sort of explanation together. One thing was obvious right away: many of the earlier puzzlements about the behaviour of liquid helium, and even Onnes's inability to reduce the temperature below about one degree, however hard the vapour above liquid helium was pumped away, had been caused simply by superfluid helium creeping up the sides of the cryostats. Other phenomena such as the fountain, were caused essentially by there being a mixture of Helium I which was not superfluid, with superfluid Helium II. In these experiments the superfluid liquid could be considered as flowing without friction through the normal Helium I. But no generally acceptable theory which accounts for all the behaviour of helium below 2·2 degrees has even yet been developed.

The war in 1939 brought this great burst of scientific discovery to a halt. And even when the war was over the restart of cryogenics was slow. We have seen that the Collins helium-liquefier, developed immediately after the end of hostilities, brought cryogenics within the reach of far more laboratories than ever could have afforded the factory-like installations of Leiden. But the great eruption of nuclear science, and the peacetime development of radar and the sciences involved in similar developments, absorbed a very large proportion of the best scientific manpower; even more important, they attracted most funds. Experiments at liquid helium temperatures, even experiments in the technologically promising field of superconductivity, remained very much 'pure science', predominantly the province of university groups.

There were, however, two developments that were later to achieve significance. First, it was shown that helium could be turned into a solid—but not by reducing temperature alone, as with all other substances. To solidify helium it was necessary to lower the temperature enough to liquefy it and then apply a pressure of at least twenty-five atmospheres (twenty-five times the pressure exerted on us by our atmosphere at normal temperature at sea level).

Secondly, there was the discovery and isolation of helium$_3$. This is an isotope of ordinary helium, which is helium$_4$. Both substances are normally gases and have exactly the same chemical behaviour at all ordinary temperatures and pressures, but whereas the ordinary helium has four particles in its nucleus (two protons and two neutrons), the rare isotope, helium$_3$, has two protons but only one neutron in its nucleus. Helium$_3$ can be found in nature, though it is very rare. But the hydrogen-bomb component, tritium, which also has three nuclear particles and is an isotope of hydrogen, decays into helium$_3$, so supplies of this material started to become available in the late 1950s and are nowadays easily obtainable and quite cheap. Helium$_3$ showed some very interesting differences from 'normal' helium$_4$. It could be liquefied, but only at a temperature at least one degree lower than the mark at which helium$_4$ liquefies. And although only tiny droplets of liquefied helium$_3$ were available in the 1950s and 1960s, it became clear that there was no 'lambda point' equivalent to the 2·2 degrees at which the behaviour of helium$_4$ changed so drastically. Above all, no one could make helium$_3$ superfluid.

But it was possible to make helium$_3$ into a solid—and because of the very strange nature of this solid there was found another method of cooling which is called 'Pomeranchuk cooling' from the name of the man who proposed it. Solid helium$_3$ is in fact less 'ordered' than liquid helium$_3$, which is exactly the opposite of the situation in all other materials. Normally a solid is more ordered than a liquid, for all the atoms are in their ordered places in the crystal lattice of the solid state, and nothing like as free to move as the atoms in a liquid. While this is true of the atoms of solid helium$_3$, it turns out that the magnetic order of solid helium$_3$ (because of the odd number of three particles in the atom's nucleus) is very much less than that of the liquid, so much less as to outweigh the extra ordering brought about by solidification. This all leads to the extraordinary effect that, under the right conditions, heating liquid helium$_3$ causes the liquid to turn into a solid. Using this effect, Pomeranchuk cooling can be achieved.

In the actual operation the helium$_3$ which has been cooled enough to be liquid is kept at a constant pressure of about thirty-two atmospheres—this is achieved by a flexible bellows operating in liquid helium$_4$. The pressure causes some of the liquid helium$_3$ to solidify, and then if the remainder of the liquid helium$_3$ is slightly heated more of it will turn solid (since the solid is of less volume than the liquid the flexible bellows reduces the volume of liquid to keep the pressure constant). We can regard this as concentrating disorder (entropy) in the increasing volume of solid $helium_3$—and since the liquid $helium_3$ is losing disorder, it is becoming more ordered and therefore cooling. This technique can be used, therefore, either to cool liquid $helium_3$ to even lower temperatures or to cool some other substance which is in contact with the $helium_3$.

Fairly elegant Pomeranchuk cooling cycles can now be used to bring the temperature of $helium_3$ down to one milli-Kelvin—that is, one one-thousandth of a degree above absolute zero.

A more easily constructed refrigeration cycle—the dilution refrigerator—was developed after Pomeranchuk cooling. The dilution refrigerator depends on the differences between $helium_3$ and $helium_4$. $Helium_4$ is a superfluid at temperatures between one degree and one-tenth of a degree above absolute zero, whereas $helium_3$ is not a superfluid in this range. So $helium_3$ can be driven through superfluid $helium_4$ almost as though it were an ordinary liquid passing through a gas. And at these temperatures a concentration of 6% of $helium_3$ will dissolve in $helium_4$—just as alcohol dissolves in water. Finally, $helium_3$ is lighter than $helium_4$. So it is possible to create a machine which works in many ways like a domestic refrigerator, in that it compresses $helium_3$ at room temperature and then cools it down to 1 degree, and uses heat exchangers, with the final production of a bath which contains pure $helium_3$ floating on top of a mixture of $helium_4$ with 6% $helium_3$. But if we look at this bath from the point of view of the $helium_3$ it is as though the pure $helium_3$ was a 'liquid' with a 'gas' of $helium_3$ (actually dissolved in $helium_4$) below it. We are therefore in the position of boiling off or evaporating $helium_3$ when we circulate $helium_3$ from the pure liquid through the mixture of two heliums—and evaporation as usual leads to cooling.

Commercially manufactured dilution refrigerators are now available which will give temperatures as low as 7 milli-Kelvin. Special models are becoming available that can reach 2 milli-Kelvin—just two-thousandths of a degree above absolute zero. With the development of these commercially available machines, a well-equipped cryogenic physics laboratory will nowadays offer its research students vacuum pumps with boosters so that they can reduce the temperature of their liquid helium to less than one degree above absolute zero and perform regular experiments at temperatures of 0·8 degrees. The main cooling apparatus of the laboratory will be dilution refrigerators taking temperatures down to a few thousandths of a degree above absolute zero. The most important work of the research group will add Pomeranchuk cooling to this, and have a final stage of nuclear magnetic cooling—similar to the device used by Simon and his team at Oxford in the early 1950s. With all this, it is possible to reach temperatures of as little as a few millionths of a degree above absolute zero in the nuclei of copper experimental specimens. But the bulk of the material is only weakly linked to the nuclei and the lowest temperature ever recorded in a bulk material surrounded by liquid $helium_3$ is 0·45 milli-Kelvin, less than half of one thousandth of a degree above absolute zero.

Once again the combination of new techniques of cooling with discoveries of new states of matter at ever lower temperatures has opened up new horizons of scientific research. Dilution refrigerators and Pomeranchuk cooling have eventually led to temperatures low enough to allow liquid $helium_3$ to turn into a superfluid. This transition occurs at a temperature of about 0·0027 degrees above absolute zero—less than three-thousandths of a degree. The original claim that superfluid liquid $helium_3$ had been found came as long ago as 1964 from the Russian team led by V. P. Peshkov of Moscow. But it was disputed by the leading American cryophysicist, J. C. Wheatley, then of Urbana, Illinois.

With remarkable similarity to the story of Josephson, it was again a young research student, working for his *Ph.D.*, who finally proved that even $helium_3$ could be turned into a superfluid. The work was done by Osheroff at Cornell University in the USA, and he achieved the effect by reaching a temperature of less than 2 milli-Kelvin, in 1972. This immediately produced a remarkable new upsurge in studies of the behaviour of matter, or at least of helium, at extremely low temperatures. The Cornell group has produced many further results; Wheatley, now at La Jolla in California, continued his role as a leader in the field of cryophysics; a Finnish group, at Helsinki, have made important contributions, and Leggett at Sussex University, in England, has emerged as probably the leading theorist dealing with the new phenomena. Many other groups, from Holland to Japan, have come into the field in the last five exciting years.

It very soon emerged that superfluid helium$_3$ has to be divided into two different substances helium$_{3A}$ and helium$_{3B}$. Both liquids are magnetic—that is, they are drawn to the poles of a strong magnet. This is in clear contrast to the virtually non-magnetic superfluid helium$_4$, though helium$_{3B}$ is clearly the less magnetic of the two. Many experiments study the properties of helium at very low temperatures by watching the behaviour of rotating discs or cylinders inside baths of the superfluids. Others rely on the behaviour of sound waves to reveal the properties of the heliums. And there are also many studies of the magnetic behaviour of helium$_{3A}$ and helium$_{3B}$.

The whole of the field is full of strange-sounding concepts. There are papers on 'magnetic super-leaks'; papers on 'second sound'. There are review articles on the 'mobile spaghetti' explanation of superfluidity in helium$_4$; and there are the concepts of 'spin waves' and 'vortices' in attempts to explain the observed phenomena.

Many of the current experiments being performed, or planned, today are attempts to find a way through this jungle of often competing ideas—in fact, to formulate satisfactory theories of superfluidity. Curiously enough, it is easier for scientists to explain the superfluidity of helium$_3$ than the same state in the much more 'normal' helium$_4$. This is because helium$_3$ has only three particles in its nucleus (two protons, one neutron), which means that the whole atom has a spin. It is therefore possible to imagine pairs of atoms going together which have the same spin—'spin up' and 'spin down' are the words used. Such pairs of atoms can then behave in motion in a way parallel to the electrical behaviour of 'paired' electrons in the BCS theory of superconductivity, and electrons, too, have a spin of their own.

Helium$_4$, however, with its four nuclear particles, has no overall spin, and cannot therefore have its superfluidity explained even partially in this way. As long ago as the 1950s an American, Feynman, produced some equations which provide answers corresponding to some of the properties of superfluid helium$_4$, but whereas the BCS theory applied to superfluid helium$_3$ predicted certain further results which have since been verified experimentally (and is therefore a useful concept in this field), no one has yet been able to make predictions from the Feynman equations that have led to illumination about the behaviour of superfluid helium$_4$.

Lacking a satisfactory theory for what they observe is, in fact, very exciting for most scientists in the field. But even more exciting is that by studying the behaviour of helium at extremely low temperatures they are able to test the laws of quantum mechanics, the basic doctrines of modern physics, in a way that has never been possible before. The laws of quantum physics were propounded in the first quarter of our century. They have commanded assent ever since, because they are the only way we have found of explaining many observed phenomena. But to be able to test the value of this assent in detail and, as one may say, at close quarters, is something rather new.

According to the great theorists there are two ways of looking at the basic behaviour of matter, based on two different statistical concepts. There are Einstein-Bose statistics, and Fermi statistics. Helium$_3$ behaves, in the last analysis, as a Fermi liquid—at extremely low temperatures its particles follow the behaviour of Fermi's statistics. Helium$_4$ is a Bose-Einstein liquid. The essential difference is that helium$_3$, having only three nuclear particles (two protons and one neutron) must have an odd number of spins, and should therefore obey the Fermi statistics; while helium$_4$, having an even number of nuclear particles, should follow the Bose-Einstein rules. Different forms of behaviour can be expected if atom pairs have their spins in the same or opposite directions. And so there comes a new and strange-sounding set of phrases used in discussing this work—there are 'up-up pairs' and 'up-down pairs', according to whether they are considering pairs of atoms with spins in the same or opposite directions.

As we have seen, there is at the moment no theory, no explanation, which covers all the results that have been observed. Such is the nature of science that one may have already emerged but it has not yet won acceptance; on the other hand, it may well be several years more before a satisfactory theory is even developed. This is the excitement of working at the frontiers of knowledge—the next theory, the next set of observations, may be the crucially important one—the winner. Or all may be thrown into confusion in the next few weeks by the appearance of a paper from someone else's laboratory which puts everything back into the melting pot again.

But the struggle to get ever nearer to the unattainable absolute zero is not a mere making of new marks, or a setting of new records for lowest temperatures. It is proving again, as it has shown throughout the hundred and fifty years of its history, that the colder scientists can make things, the more interesting are the states of matter that are revealed.

10 The Future

By definition it is impossible to predict what science will discover. More to the point, there would be no fun, no intellectual interest, in being a scientist if you knew what you were going to prove. Any attempt, then, to predict what will emerge from the continued study of very low temperatures in the future can only be foolhardy. It is, however, possible to see some trends emerging which will surely continue over the next few years. It is also possible at least to guess what may be some of the interesting questions that scientists of the future will be asking. It is also possible to see the directions in which the technology of the very cold is heading, though here again there will undoubtedly be twists and turns which no one can reasonably foresee. And finally it is possible to point the finger at one development, which many people hope and wish to see take place, yet which even our present knowledge is sufficient to identify as a mirage.

For the pure science of cryophysics, the main obstacle at the moment is the sheer tactical difficulty of carrying out the experiments and obtaining the readings or making the observations. The cooling apparatus, the basic necessity for getting into the temperature regions or states of matter that are the object of study, is really rather large. Several stages are necessarily involved—the provision of liquid helium with its attendant insulated cryostats; a dilution refrigerator; a stage of Pomeranchuk cooling, and perhaps finally a magnetic cooling. The final specimen or example of material is perforce very small and it is totally enclosed by machinery. The electronics necessary to obtain the readings are fairly sophisticated and, of course, they do not extract from within the cooling machinery anything as simple as a direct reading of some quantity; they are more likely to present some continuously varying wave form on a cathode ray oscilloscope. This result must then be interpreted in terms of the behaviour of the superfluid helium within.

It is therefore reasonable to expect that in the years immediately ahead and perhaps into the early 1980s the main scientific effort in the field of very low temperature research will be directed towards getting technical advances in the experimental equipment, so that more results and readings can be obtained more easily and more reproducibly. The broader basis of observational results will in turn help the theoreticians, and we may certainly expect a coherent and acceptable theory of superfluidity to emerge shortly.

There is at the moment no major new method in sight of cooling or for achieving even lower temperatures than the present partial penetration into the range of a few millionths of a degree above absolute zero. It is, however, unlikely, in view of the general history of science, that nothing further will turn up. There are proposals being mentioned for the use of lasers for cooling—the theory being that the great orderliness of laser-light could be used to impose even further orderliness on other materials.

The most exciting results in low-temperature physics in the near future are likely to come as other materials are cooled down to the temperatures that are now only reached by liquid helium and copper. What these results will be it is impossible to tell. But a hint was given in April 1977 with the discovery, again by Bernd Matthias, of the University of California—Bell Laboratories, that a rare material, a compound of boron, erbium and rhodium, which is definitely superconducting at 8 degrees above absolute zero, ceases to be superconducting when its temperature is brought down even further, to below one degree. In fact, it appears to become magnetic at this low temperature.

Certain parallels to this unexpected finding have previously been found in non-superconducting materials when brought down to the temperature of liquid helium or just below. This has been called the Kondo effect. Measurements of the electrical resistance of these materials showed the expected steady drop as temperature was lowered, but then instead of going superconductive, as happens in many cases, the resistance of the material started to increase again as the temperature was reduced even further. The Kondo effect has been established as closely related to the precise quantity and nature of impurities in these materials, but since it is virtually impossible to produce completely pure materials, there could well be further developments in this field, both theoretical and possibly practical, for most modern electronic devices of the transistor family depend on the accurate and careful introduction of 'impurities'.

In all these areas there are exciting possibilities for 'pure' or 'basic' scientific research, and certain progress to be made. The exact nature of the advances is hidden from us by the nature of science itself. There is however one area where one can confidently predict that there will be no progress on any logical basis—probably much sorrow and heart-searching. This sad, and possibly serious, development of low-temperature science is the deep-freezing of human corpses by what are sometimes termed 'cryonic societies'. The object of the operation is to preserve the body intact for some period of years—as long as thirty years has been mentioned—in the hope that in that time medical science will have discovered a cure for the disease from which the person died. Then the body will be thawed out and rescuscitated and cured of the disease. So the sufferer may be able to live again. The notion of keeping human bodies at liquid-nitrogen temperatures, in the same way that sperm and blood cells are perfectly satisfactorily stored for later resuscitation and use, originated in the United States and is most practised there. But it is reported to have spread to France, Belgium and Britain since then. By 1975 it was believed that as many as thirty or forty corpses were so stored in deep-frozen conditions. The most clearly outspoken condemnation of the practice came from Professor Nicholas Kurti, one of Simon's original Oxford team of cryophysicists and himself a pioneer in the field of superconductivity. In 1975 he spoke at the Jubilee Assembly of the Hungarian Academy of Science, celebrating its hundred and fiftieth anniversary, and he concluded his speech by saying this of the cryonic societies. 'One may laugh this off, but the sad thing about it is that relatives probably honestly believe that their beloved will be resuscitated, and to give hope where no hope exists is very cruel indeed. The fact is that several hours elapse between death and the freezing process and, as is known, the brain suffers irreversible damage if the blood supply ceases for more than a few minutes. A successful resuscitation would only be possible if a method could be developed to perfuse the whole body with a protective agent before death so that no damage is sustained during the cooling, and then to remove it during rewarming. But I firmly believe that today and for many years to come this sort of speculation and all ideas about prolonged suspended animation fall in the realm of science fiction and not of serious science.'

The subject is rather more sympathetically treated by Robert W. Prehoda (1969). This book makes what will probably in the long term be the most important point, that there is much to be learned from further studies of the mechanisms of hibernation in animals and insects. Natural hibernation is, indeed, a rather neglected field of research, despite the fact that there are some very unexpected physiological mechanisms at work when animals go into the state of dormancy, and it is indeed reasonable to expect that valuable knowledge might emerge from further studies in this field.

But even before Prehoda's book appeared, scientists were looking in this direction and found themselves unable to be optimistic. 'The whole concept of suspended animation, the interruption of time, is an exciting one and holds the same emotional charge that is found in such other examples of science fiction come true as space travel and energy from the atom. Herein lies one of the hazards to cryobiology. The fact that individual cells can be preserved by freezing carries the implication that tissues, organs and even whole animals can be similarly suspended. However, as should be evident from the many reports cited in this book, not all cells respond equally well to freezing. In particular not all cells respond equally well to any single freezing procedure. For some species the percentage recovery is very small.'

That warning by Harold T. Meryman, already quoted in Chapter 6 of this book, is worth repeating. It was written in 1966; it remains true today. The most exciting developments of low-temperature science, the developments most appealing to the human imagination (perhaps because of our sense of self-preservation), lie not in the exotic physics and chemistry of the approaches to absolute zero, but in the advances of the biosciences, operating mainly at temperatures down only to minus 196 degrees, the temperature of liquid nitrogen.

But after the extraordinary early advances of cryobiology, the freezing of sperm, blood cells and simple tissues, Meryman's words of caution have proved all too wise. Progress, considerable progress, has undoubtedly been achieved (see Chapter 6), but the bigger visions have not yet materialized as practical possibilities—yet. It is still impossible to store a living kidney for more than a few hours. It is now practical to transport an organ such as a kidney from one hospital where the donor has died to another where a potential recipient is waiting, as long as the storage time is no more than eight to ten hours at most. But no more than this time lag can be accepted for human medical purposes. Even the experimentalists are setting themselves a target of storage for no more than seventy-two hours, to be followed by a successful transplant. The days of 'organ banks', storing hearts, kidneys and livers so that a patient needing a transplant can be offered an organ more or less perfectly matched to his own tissue-type, are still very far away.

On the other hand, those engaged in clinical research in cryobiology feel that they are making steady progress at least towards the storage of kidneys for a reasonable time. They believe that there are no obstacles of principle, and that their objectives can be achieved by continuous technical advance in the use of perfusates (liquids, probably chilled, to wash out and replace the blood from the kidney) and in the procedures of cooling and cold-storage. In theory, the heart should be easier to store than the kidney, and in practice it seems that liver transplants face fewer difficulties and rejections than kidney transplants, but so far there has been little incentive to work on the storage of hearts and livers since the transplant operations themselves are so rare.

If Harold Meryman's cautionary advice should prove to be more applicable to the freezing and storage of major organs than we hope, then it may well be that further advances in cryomedicine will come from the freezing and storage of single cells and small pieces of fairly homogeneous tissue. There is, for instance, a very considerable increase in the number of bone-marrow transplants in modern medicine and this sort of transplant is being used in an ever-widening number of diseases; it is a technique that is no longer being used only for the treatment of children born with bone marrow deficiencies, but is also being used, or considered for use, in cases of some anaemias and leukaemias in conjunction with other treatments. At present the transplants are taken 'live' from the donors—the procedure is more akin to a complicated blood transfusion than to the dramatic removal of a major organ from a living relative or a cadaver. But it is not impossible to visualize the day when short-term storage of the marrow becomes necessary and perhaps eventually there might be banks of specialized bone-marrow cells.

Rather further into the future is the possibility that transplants of the pancreas, or at least of pancreatic tissue, may become a treatment for diabetes. Certainly transplants of this tissue have already taken place. Diabetes is such a common condition that if this technique does represent a major development in medical practice there may come a time when storage of pancreatic tissue could be important.

But it is more likely that cryomedicine will develop most strongly from further applications of the techniques of cryosurgery and from advances made in centres such as the European Blood Bank in the freezing and storage of new or rare types of blood components. Less dramatic than the vision of organ banks waiting to supply us with a fresh heart or liver when we suffer a major crisis, this is nevertheless a safer, wiser and more conservative prophecy to make.

Scientifically, the important achievements in those fields of cryobiology applicable to animal breeding have probably already been made. In this field the future will surely see a steady and ever wider application of the techniques of storing fertilized eggs (ova) and their re-implantation into foster mothers. These techniques will be applied, as the pioneering experiments have projected, to the breeding of cattle and farm animals of all sorts, as well as to laboratory animals such as mice, rats, rabbits and guinea-pigs.

It is also likely to be a safe prediction that the techniques of freeze-drying bacteria and other micro-organisms will be another major application of cryobiology in the future. We tend to regard micro-organisms as 'germs', pathogens threatening our health; but man has long used micro-organisms for his own purposes, notably in the making of cheeses and in brewing. A major tendency in our own time is to use micro-organisms more and more consciously. Some are bred and cultured and harvested to produce penicillin and the other antibiotics; in the last few years others have been 'domesticated' to produce protein from oil and gas, and this food product has gone on to the market as cattle food in many countries of the world; we are now beginning to use micro-organisms for mining metals, using the creatures' power of concentrating the metal atoms in their bodies; if 'genetic engineering' lives up to its promise we shall be using bacteria and yeasts and fungi from the soil and from our own bodies to make medicines and industrial chemicals and fertilizers for our fields. All these techniques of domesticating and using micro-organisms will require the storage and transporting of special strains of bacteria or yeasts, and the establishment of many reference collections, or banks of micro-organisms. These can only be achieved practicably by use of freeze-drying techniques pioneered by the cryobiologists.

It appears to be a law of nature, never enunciated by any scientist but secretly acknowledged by most writers, be they historians or mere journalists, that it is those things which are least dramatic which most affect most people. And it seems likely that this law holds in the field of predicting the future of low-temperature work. More of us will probably be more affected, and better pleased, by comparatively minor advances in food storage and treatment techniques at undramatic temperatures between 0 and minus 30 degrees than by exotic medical and surgical techniques which will always be applicable only to a few sufferers.

Huge quantities of the world's harvests are destroyed by pests of all sorts while they are awaiting consumption. And this applies most in those very areas, the countries of the Third World, where the preservation of crops is most important to people who have little enough to eat in any case.

The whole object of domestic refrigerators is in fact to stop spoilage, to protect our harvested food from damage by bacteria through reducing the temperature and thus reducing the activity of the micro-organisms. Plainly, it is most unlikely that Western-style refrigerators or deep-freezers will spread on a large scale to the developing nations in the near future. Nor is it likely that refrigerated storage of grain crops in India or Africa will prove economical or even practicable, for although refrigeration might be protective against the pests of grain or rice stores, it is not the normal method of providing such protection.

But almost certainly, in the opinion of refrigerating industry experts, there will be large and important developments in cold storage in the Western or industrialized world. Many of these will be purely technical, enabling us to achieve our present-day capacities for removing heat and maintaining cold with far less expenditure of energy. It is worth re-emphasizing that our modern city life is totally dependent on our power to refrigerate. Our cities have grown too big for their inhabitants to be fed with fresh foods unless we can keep these foods 'fresh' by refrigeration; the supermarkets and even small local shops, the whole system of distributing food depends on refrigeration.

It is likely that the use of cooled or refrigerated storage will increase vastly. If we can develop low-temperature storage for the wide variety of fresh vegetables, fresh fruits, salad stuffs, and even bread, cakes and confectionery, we can cut out huge amounts of food wastage and make our food supply and distribution systems cheaper and more economic. It is felt that this development will provide the largest expansion in refrigeration techniques and sales in the medium-term future.

The fastest-growing application of cryogenic engineering is undoubtedly to be found in the liquefied natural gas (LNG) business. As oilfields are developed in more and more exotic or difficult areas—in Indonesia and Angola, in Alaska, the North Sea, and off the Australian West Coast—it becomes more and more important to develop liquefaction facilities so that the natural gas associated with the oil can be transported with reasonable economy to the centres of industrial production and the big centres of Western urban life. Just as the Arab countries of the Middle East have come to insist that their gas should not just be flared away, so we can see the British Government actually holding up production of its precious North Sea oil rather than allow gas to be flared off and wasted. The network of pipelines for gas is growing, crossing the Mediterranean and the North Sea, for instance. But for truly long-haul operations, such as from Indonesia to Japan, only transport of refrigerated and liquefied gas seems possible.

Meanwhile the industrial gas industry is continuing to expand and can be expected to continue to do so. The drive to sell liquid nitrogen has now resulted in its becoming a major product of the industry, rather than a somewhat embarrassing by-product. New applications for liquid nitrogen are continuing to be found. One of the most interesting is the British Oxygen Company's development of a cryogenic process for hardening the metals used in machine tools. The cutting components of machine tools are invariably made from particularly tough alloys of steel. They are made, therefore by a heat-process and are finished with further heat-treatments to render them tough, for the life of the cutting component of a machine tool, and the necessity for constant regrinding to sharpen the edges are major factors in the economics of engineering production processes. But now BOC has shown that gradual cooling of a heat-treated machine tool component and then lengthy storage at low temperatures can double the working life of the component. The cooling is achieved by spraying with liquid nitrogen, and then the specimen is immersed for several hours in a bath of liquid nitrogen at minus 196 degrees. Thawing is achieved simply by taking the component out of the nitrogen and allowing it to warm up to room temperature naturally.

What makes this development particularly interesting is that no one knows why the process works and has the effect of toughening metal. All that is known so far is that the extreme cooling penetrates right to the centre of the component, so the effect cannot merely be a surface hardening. This is confirmed by the fact that the metal remains tougher than normal even when it does finally need regrinding.

All predictions of industrial and technological growth in the last quarter of the twentieth century are bedevilled, however, by the uncertainties engendered by the threat of an energy crisis. According to recent studies, the crisis, sometimes called the energy gap, could appear as early as 1985, by which date it is estimated that overall world demand for oil and natural gas will begin to outstrip production. Whether or not there then begins an actual oil shortage is a matter of much dispute. What seems reasonably certain is that there will be a rise in the price of oil of such magnitude that our present system of costings and economics will be severely disrupted. The temporary shortfall in oil production caused by the Iranian revolution in 1979 with its immediate effects on costs merely emphasizes the longer-term dangers. Since reserves of oil must be finite while demands for energy can theoretically be infinite, there will inevitably come a time in the foreseeable future when we will have to limit the use of oil. First we shall have to stop burning oil for heating and electricity production—we shall simply not be able to afford this most wasteful use of fuel. Later we may have to develop modes of transport that are not dependent on oil, so that the last reserves of hydrocarbons can be used as raw materials for medicines, plastics, artificial fibres and similar essentials.

There are many possible ways of meeting this coming change in our industrial and technological system. One in particular will heavily involve the industrial gas industry and the techniques of cryogenic engineering. This development—which is at the moment purely theoretical—is called 'the hydrogen economy'. In principle the hydrogen economy looks to liquid hydrogen as the transportable fuel of the future. The hydrogen itself will be extracted from water using power produced by nuclear fission or nuclear fusion electricity-generating stations. And since liquid hydrogen transport, storage and use will almost certainly need cryogenic techniques, it follows that we will have low-temperature systems in our cars and lorries, in our houses and offices too.

Liquid hydrogen has shown its value as a fuel in the space programme. It was first used on a large scale in the American Centaur rocket stage. It has been used successfully in the Apollo programme and continues to be used in the launching of heavy satellites or distant space probes by the Titan/Centaur launch vehicles. It is assumed that the Russian space programme probably uses liquid hydrogen, too. A wide public has become familiar with the outward signs and appearances of large cryogas systems, with their big insulated tanks, heavy feedlines and jets of escaping and boiling gas appearing like pure white steam in all the television pictures of rocket launches from Cape Canaveral.

The hydrogen economy would see the big electricity power-stations (using the released energies of nuclear atomic processes as 'fuel') operating throughout the night-time hours, when industrial and domestic demand for electricity is low, to separate hydrogen from water in the process known as electrolysis. The hydrogen would be pumped round the industrial system probably by the pipeline grids at present used for transporting natural gas. At many centres it would be concentrated, liquefied and stored in gigantic insulated Dewar flasks. This, of course, is where the cryo-engineering would start. But the techniques of dealing with extremely cold gases, and the materials necessary for handling them safely, would expand outwards from these central storage points into every corner of our lives, for the hydrogen would have to be used in liquid form, certainly in our cars, and probably in our homes as well. The roadside petrol-pump attendant would have to learn the safety rules for handling liquid hydrogen, just as he now observes the rules for handling petrol.

The most likely way of using the hydrogen to drive vehicles or provide other sources of local energy is through the fuel cell, which reunites the hydrogen with oxygen to reform water, producing an electric current in the process. Fuel cells have been used extensively by the astronauts to provide both power and pure water in the Apollo spacecraft, with the oxygen and hydrogen stored in liquid form in separate tanks. (It was the disruption of one of these tanks by an internal explosion that caused all the trouble in the near-catastrophe of the Apollo XIII mission, from which the astronauts narrowly escaped with their lives.)

There are, however, variations upon this idea. The Euratom laboratory at Ispra in Northern Italy is probably Europe's most important centre of hydrogen research, and the scientists there have worked out alternative methods of obtaining large quantities of hydrogen from a series of chemical reaction processes. At the other end of the hydrogen chain, experiments in America show that existing motor-car engines can work quite satisfactorily using liquid hydrogen as fuel. One of the greatest advantages of the hydrogen economy is that the burning of hydrogen as fuel gives nothing but water, and is therefore completely pollution-free.

The greatest impact of low-temperature science and engineering on our world in the next decades will undoubtedly come from the application of superconductivity to more and more spheres of industry, power-engineering, electronics, military programmes and even to medicine. It is superconductivity that will bring liquid helium, the concept of temperatures 250 degrees and more below freezing point, with all the associations of deep refrigeration and insulation, to the knowledge and experience of vast masses of people.

'Superconductivity is the finest tool ever to fall into the hands of electrical engineers. I have no doubt Nature intends us to use it.' These were the words of Mr A. D. Appleton, of the International Research and Development Company of Newcastle-upon-Tyne, the man in charge of building the Fawley motor, the world's first large superconducting motor. He used the words in the final section of his Hunter Memorial Address, given to the Institution of Electrical Engineers in London in 1976. We must remember, too, the words of the American, Dr Henry Kolm: 'Superconductivity is beyond question the most significant technological innovation since the invention of the wheel. . . . The wheel provided us with frictionless transport of matter and superconductivity provides us with frictionless transport of electricity. It is every bit as important as the transport of matter. In fact, the wheel has enjoyed a period of usefulness which is rapidly coming to an end, considering our highways and countryside, whereas superconductivity has not begun yet.' That was in a television programme in 1974.

We have seen (in Chapter 7) that superconductivity has begun to invade the world of practical engineering since then, by winning its first really large-scale job in providing the magnets for the Energy Doubler/Saver in the world's largest nuclear accelerator at Fermilab in Illinois. And small super-conducting magnets are now regularly and commercially on sale as part of scientific machines such as nuclear magnetic resonance spectrometers.

It is generally accepted that superconducting magnets of very large size will be necessary if we start to build large power-stations generating electricity from controlled nuclear fusion. Such building will probably occur around the start of the next century—in some twenty years' time. It seems that only superconducting magnets will be able to generate economically the very powerful magnetic fields which will be needed to control the very high temperature gas-plasmas which are needed for energy-producing fusion reactions—reactions which are essentially controlled versions of hydrogen-bomb explosions.

Small superconducting magnets will always suffer the drawback of the cost and size of the liquid helium containers, insulators and refrigeration equipment, though they will doubtless slowly creep down-market from their exotic scientific uses into the more sophisticated and small-scale industrial processes.

But where will superconductivity start to come into the ken of the common man? Few doubt that it will do so within the next two decades and there are four main possibilities: in the generation of electricity, in the transmission of electricity, in the provision of large motors for ship and similar propulsion, or in the host of possible devices that are grouped under the title of 'superconducting electronics'. In one or all of these areas, superconductivity will soon begin to affect our everyday lives.

The trend towards applying superconductivity certainly seems strong in the electrical generating industry. A large body of basic research work is accumulating in both industry and university laboratories. No insuperable obstacles have appeared in the course of this, and much work has gone into the practical problems such as checking the behaviour of liquid helium under the forces of rotation that it would undergo in large electrical generators. It seems unlikely that the effort expended by establishments like the Massachusetts Institute of Technology or Southampton University, or the investment of a company as large as Westinghouse, will go without some result. But even more convincing are the sort of arguments in favour of the superconducting machines, arguments that fall in with current trends in thinking. The arguments in favour of such machines are not just in terms of dollars or pounds or probable cost-effectiveness—we know too well that such predictions all too often have turned sour when the full cost of development of a new technology are finally worked out. Superconducting machines recommend themselves because they are smaller, easier to handle, easier to fabricate. There are economies of large size in most engineering fields, but there is a feeling that in power engineering, as in many other areas of the energy-using field, we are reaching the limits, the social if not the engineering limits, of very large units. The trend towards large single-unit turbo-alternators for generating electricity has not lived up to its original promise, largely because unforeseen problems such as the difficulties in fabricating the very large shafts and the strains produced by supersonic travel speeds in the turbine blades, have cropped up. The superconducting generator offers a way out of the possible blind alley of very large machines, as well as offering clear-looking promises of good economics, because, despite the cost of the necessary refrigeration, it has basic design advantages such as higher power-density.

Superconducting transmission of electric power, despite its attractions of virtually no resistance to the passage of current and therefore the very low loss, has fallen from its original state of high promise. The gas-filled cable offers such an attractive improvement upon present standards that the necessity for anything further, such as the superconductive cable, has been pushed into the more distant future. And even then, cables of aluminium, not superconductive but cooled to liquid nitrogen or liquid hydrogen temperatures, may always prove more economical than the high-cost materials and the liquid helium temperatures of the superconductor. It will be a strange irony if the last thing we use superconductivity for is conducting electric power.

The difficulty in introducing the direct-current superconducting machines is rather different; it is the need for an incentive. Superconducting direct-current motors have been shown to work in practice, and so they are an important step ahead of the alternating-current machines being developed for power generation. Obvious places where the superconducting motor should find a useful role include ship propulsion, providing motive power for large steel rolling-mills, or turning the winding gear in very large, deep mines—anywhere, in fact, where high power output is needed at comparatively low speeds. But who will take the plunge and order the first? It seems unlikely to come from the commercial shipbuilding industry where there is so much unused production capacity around the world, and where the emphasis is on commercially competitive prices in a cut-throat market, so that innovation is hardly encouraged. It still seems most likely that the pump must be primed by governments and that the superconducting motor is most likely to make its first serious impact in warship propulsion units.

Superconducting electronics will also be most likely to make an impact on our lives through the intervention of the military/industrial complex. And unless one can visualize some miniaturization of the liquid helium refrigeration unit, superconducting electronics are never going to arrive in our living-rooms or kitchens as the ubiquitous transistor of normal-temperature electronics has done. But SCE will surely arrive in hospitals, the medical electronics industry and the computer world.

In virtually any direction we look, we see more applications of cold ahead of us. And in all these possible future developments, the advance seems to lie in the direction pointed by the phrase, 'the colder the better'.

References, Acknowledgements and Notes

In the hope that it will be more convenient for any reader wishing to follow up details mentioned in this book, I have grouped the references to papers, articles and similar sources by subject matter, rather than listing them in the precise order in which they occur in the book. Under each of these subjects I have also mentioned those people who helped me to try to understand their work. I hope that this will show my gratitude every bit as well as listing them in alphabetical order in the foreword or introduction.

HISTORY OF REFRIGERATION

Gosney, W. B. (1968). 'Modern Refrigeration'. *Journal of the Royal Society of Arts*, May, p. 448. This contains his three Cantor Lectures entitled 'Principles of Mechanical Refrigeration', 'The Preservation of Food by Cold', and 'Applications of Refrigeration'.

Gosney, W. B. (1973). 'A Pioneer Voyage'. *Meat Trades Journal*, 8th November.

Locke, Geoffrey (1975). *Ice Houses*, included with the Annual Report. The National Trust, London.

Oldham, Bernard C. (1969). *Cold in the Service of Man.* Paper read before the Institution of Mechanical Engineers, London. Awarded the Institution's Silver Medal.

Smith, Edgar C. (1943). 'Historical Record of Refrigeration'. *Modern Refrigeration*, 15th April, 79.

The Institute of Refrigeration published a short history of its seventy-five years of existence, which contains further general information on the subject. I am grateful to the Institute's Secretary for his help in providing this material and for other advice about the refrigeration industry. Mr James Douglas, a prominent member of the industry and Chairman of the BRACA Education and Training Section, also provided help and information. Dr Gosney spent a great deal of time elaborating on his Cantor Lectures, expanding on them and explaining further points.

FOOD PRESERVATION

Some of the references for the first subject plainly overlap into this section—notably Dr Gosney's Cantor Lectures. I received help also from the Food Freezer and Refrigerator Council, and from the Ice-Cream Federation Ltd in a short paper, 'All About Ice-Cream'. T. Wall and Sons (Ice-Cream) Ltd supplied further information through their Information Officer.

The Torry Research Station of the Ministry of Agriculture, Fisheries and Food at Aberdeen, through the Head of the Information Service, Mr J. J. Waterman, made available to me eight of the *Torry Advisory Notes*, covering various aspects of freezing and super-chilling fish, treating fish at sea and transport of fish, as well as Torry Research Report No. 1, by G. C. Eddie, *Freezing on Board*. May 1961.

INDUSTRIAL GAS INDUSTRY

Flynn, T. M. and Birmingham, B. W. (1969). 'Cryogenics in United States National Programmes'. *Cryogenics*, 9, 1, 3.

Lafaurie, Michel (1968). 'Les multiples applications des fluides cryogéniques dans l'industrie et les laboratoires'. *Chimie et Industrie—Génie Chimique*, 99, 5.

The Production of Oxygen, Nitrogen and the Inert Gases, Revised edition, March, 1974. British Oxygen Company.

Scott, R. B. (1963). *Cryogenics and Space Technology*. Paper read at the XIth International Congress on Refrigeration, Munich, Germany, 28th August.

Scott, R. B. (1966). *Liquid Helium Technology at the National Bureau of Standards*. Paper read at the IIR Commission Meeting, University of Colorado, 16th June.

The Theory of Air Separation. Air Products Ltd.

The British Cryogenics Council has produced much information, including a small pamphlet, *Cryogenics*, and I am grateful to Dr R. G. Scurlock, of the University of Southampton, for making it available to me. Dr Scurlock also gave much time to talking to me about industrial cryogenics and his research in this field. Naturally, much of the information in this area came from the large industrial companies, and all that I asked were most generous with papers and information.

I have to thank particularly Monsieur Simon, Chef des Relations Extérieures of L'Air Liquide in Paris, and his counterparts, Mr Tim Furniss of Air Products, and Mr M. J. Webb of the British Oxygen Company. From these sources came much of the detailed information about various applications of industrial gases to engineering processes and the latest progress in food preservation and industrial food preparation techniques. There are two specific papers in these fields:

Hall, Ross (1975). *Shrink Fitting with Liquid Nitrogen.* British Oxygen Company, Quality Assurance News.

Pyke, Magnus (1975). *Cook Freeze and Britain's Social Future.* Air Products Ltd.

CRYO-ENGINEERING

Howard, John P. (1971). 'Supercold Industry Hots Up'. *New Scientist*, February, **49**, 737.

Perry, E. J. (1974). *The Role of Refrigeration in Energy Conservation.* Institute of Refrigeration, London.

Poole, R. R. (1974). *Ground Freezing for Civil Engineering Applications.* Institute of Refrigeration, London.

Scott, R. B. (1961). *Recent Progress in Cryogenic Engineering.* Paper read at the Mountain States Navy Research and Development Clinic, Raton, New Mexico.

CERN Courier, February 1976, p. 51, provides details of the liquid neon transport operation in the USA.

CRYOBIOLOGY

Parkes, A. S. (1956). 'Preservation of Living Cells and Tissues at Low Temperatures'. *Proceedings of Third International Congress on Animal Reproduction*, Cambridge.

Polge, C., Smith, A. U. and Parkes, A. S. (1949). 'Revival of Spermatozoa After Vitrification and Dehydration at Low Temperatures'. *Nature*, **164**, 666.

Whittingham, D. G. (1974). 'The Viability of Frozen-thawed Mouse Blastocysts'. *Journal of Reproduction and Fertility*, **37**, 159.

Whittingham, D. G. (1976). 'Low Temperature Storage of Mammalian Embryos'. Included in Workshop on Basic Aspects of Freeze Preservation of Mouse Strains held at Jackson Laboratory, Bar Harbour. Gustav Fischer Verlag, Stuttgart.

Whittingham, D. G. (1974). 'Embryo Banks in the Future of Developmental Genetics'. Included in *Symposium of Developmental Genetics.* XIIIth International Congress on Genetics. Reprinted *Genetics*, 78, 395.

Dr D. G. Whittingham also gave me an unpublished paper, 'General Aspects of Egg Culture and Preservation'; Dr David Pegg let me have a copy of a paper in preparation, 'Long Term Preservation of Cells and Tissues'. Dr Whittingham, Dr Pegg and Dr Farrant gave me many hours of their time.

I was permitted to attend the XIth Annual Meeting of the Society for Cryobiology in London in 1974, and the Meeting of the Society for Low-Temperature Biology in Amsterdam in April 1975. Much information contained in this book, and most of the information about cryosurgery, came from these sources.

Dr Offerijns showed me round the Central Laboratory of the Netherlands Red Cross Blood Transfusion Service, and explained the workings of the European Blood Bank, for which I am most grateful.

Dr Audrey Smith and Dr Chris Polge reminisced about the early days of cryobiology as well as pointing out likely developments, and it is with great pleasure that I remember my visit to them.

Professor J. R. Batchelor explained the cryogenic work on skin storage at East Grinstead.

Information on commercial sperm-freezing and animal-breeding came from the Milk Marketing Board through the kindness of Mr G. F. Smith, Head of Veterinary Services and Research, and from Dr Kevin O'Connor, Chief Scientist to the Board.

CRYOPHYSICS

For the history of early cryophysics I have to thank the Kammerlingh Onnes Laboratory at Leiden, its Director at that time, Professor C. J. Gorter, and particularly his colleagues who showed me round and explained the present work of the Laboratory. My thanks also to the Director of the Museum for the History of Science, also in Leiden. Professor Nicholas Kurti at the Clarendon Laboratory in Oxford gave me much time (as he has always been willing to help inquiring journalists). I have also used his *Memoir of Sir Francis Simon* (1958) in the series of Royal Society Biographies of Fellows.

SUPERCONDUCTIVITY

Appleton, A. D. (1976). Hunter Memorial Lecture at the Institution of Electrical Engineers, London.

British Broadcasting Corporation (1974). 'The Greatest Advance since the Wheel'. Broadcast on BBC TV as part of the *Horizon* series, 25th November.

CERN Courier (1975). **15**, 9, 27.

Essman, U. and Trauble, M. (1971). 'The Magnetic Structure of Superconductors'. *Scientific American*, March, 75.

Kurti, N., 8 May 1975, Remarks at the Closing Session of the 150th Anniversary Assembly of the Hungarian Academy of Sciences, Budapest.

Lee, J. A., Old, C. F. and Larbalastier, D. 'Some Aspects of Multifilamentary Nb_3Sn Production'. 'Physiques sous Champs Magnetiques Intenses'. *Colloques Internationaux CNRS*, No. 242.

Maddock, B. J. and Male, J. C. (1976). 'Superconducting Cables for a.c. Power Transmission'. *CEGB Research*, 4, 11.

Matthias, B. T. (1972). 'High Temperature Superconductors'. *Science*, **175**, 1465.

Scurlock, R. G. (1974). 'Superconductors'. *Electronics and Power*, 11th July, 524.

Stuckey, William K. (1974). 'Supercool Scientist'. *U.S.A. Horizons*, **10**, 22.

Superconducting Magnet Technology, 4th edition (1972). Rutherford Laboratory, Didcot, Berkshire.

Wilson, M. N. (1970). 'Filamentary Superconducting Composites'. *Composites*, December, 341.

Wilson, M. N. et al (1972). *AC3—A prototype Superconducting Synchroton Magnet.* Rutherford Laboratory, Didcot, Berkshire.

Here again I am particularly grateful to Professor N. Kurti for his time and patience with me. I have also to thank the Ministry of Defence for information about their research on superconducting ship machinery. And I must thank Dr John Hulme, Scientific Attaché at the American Embassy in London, for his great help in describing his own work with Westinghouse Corporation. Dr Martin Wilson and his colleagues at the Rutherford Laboratory were most generous with their time at my request.

SUPERFLUIDITY

The main sources for this subject were long conversations with Professor Dobbs, of Bedford College, University of London, and Dr Truscott and Professor Brewer of the University of Sussex.

The following articles in *Nature* were also valuable:

Goldschvartz, J. M. (1977). 'Magnetic Superleaks'. *Nature*, **266**, 824.

McClintock, P. V. E. (1975). 'Persistent superfluid flow'. *Nature*, **253**, 93.

McClintock, P. V. E. (1976a). 'Superfluid Helium$_3$: An impediment removed'. *Nature*, 259, 269.

McClintock, P. V. E. (1976b). 'Superfluid Helium$_3$: Magnetic pendula'. *Nature*, 259, 445.

McClintock, P. V. E. (1976c). 'Mobile Spaghetti'. *Nature*, **260**, 484.

McClintock, P. V. E. (1977). 'Spin Waves in Superfluid Helium$_3$'. *Nature*, **266**, 68.

Kendall, J. (1974). 'Helium$_3$: the new Superfluid'. *New Scientist*, **64**, 918, 100.

SUPERCONDUCTING ELECTRONICS (Josephson effects)

Broers, A. N., Molzen, W. W., Cuomo, J. J. and Wittels, N. D. (1976). 'Electron-beam Fabrication of 80–Angstrom Metal Structures'. *Applied Physics Letters*, **29**, 9, 596.

Cohen, D. (1975). 'Magnetic Fields of the Human Body'. *Physics Today*, August, 34.

Esaki, L., Giaever, I. and Josephson, B. D. (1974). The Nobel Lectures in Physics, 1973, by Leo Esaki, Ivar Giaever and Brian Josephson. Reprinted in *Reviews of Modern Physics*, **46**, 2.

Josephson, B. D. (1974). 'The Discovery of Tunnelling Supercurrents'. *Science*, **184**, 4136, 527.

Josephson, B. D. (1966). 'Superconductors, Superfluids and Wave Theory'. *Discovery*, July, 19.

Josephson, B. D. (1966). 'New Superconducting Devices'. *Wireless World*, October, 484.

Zimmerman, J. E. (1972). 'Josephson effect devices and low-frequency field sensing'. *Cryogenics*, February, 19.

I must express great gratitude to Professor Brian Josephson himself and to Dr Terry Clarke of the University of Sussex for long discussions on SCE.

FUTURES

Hampson, P. J., Hart, A. B., Jobes, B., Swift-Hook, D., Syrett, J. J. and Wright, J. K. (1975). 'Can hydrogen transmission replace electricity?' *CEGB Research*, **2**, 4.

Editorial (1973). 'Looking at the Hydrogen Economy'. *Nature*, **243**, 184.

Taylor, Dick (1976). 'Kondo—the Physicist's toy'. *New Scientist*, **70**, 1003, 513.

Bibliography

Booth, N. (1973). *Industrial Gases.* Pergamon Press, Oxford.

Critchell, J. T. and Raymond, J. (1969). *A History of the Frozen Meat Trade.* Dawsons, Pall Mall, London (new edition of original published in 1912 by Constable, London).

Cummings, R. O. (1949). *The American Ice Harvests 1800–1918.* University of California Press, Berkeley.

Dugdale, John Sidney (1966). *Entropy and Low Temperature Physics.* Hutchinson, London.

Fishlock, David (Ed.) (1969). *Guide to Superconductivity.* Macdonald, London.

Haselden, Geoffrey G. (1971). *Cryogenic Fundamentals.* Academic Press, London.

Landsborough Thomson, A. (1973). *Half a Century of Medical Research,* Vol. 1. HMSO, London.

Light, S. H. (1965). *Therapeutic Heat and Cold,* 2nd edition. Waverley Press, Baltimore.

Lom, Walter (1974). *Liquefied Natural Gas.* Applied Science Publishers, London.

Luyet, B. J. and Gehenio, P. M. (1940). *Life and Death at Low Temperatures.* St Louis University and Biodynamica, Missouri.

Masters, Thomas (1844). *The Ice Book.* Simpkin, Marshall and Co., London.

Meetham, A. R. (1967). *Depth of Cold.* English Universities Press, London.

Mendelssohn, K. (1966). *The Quest for Absolute Zero.* Weidenfeld and Nicolson, London.

Meryman, H. T. (Ed.) (1966). *Cryobiology.* Academic Press, New York.

Parkes, P. and Smith, A. U. (1960). *Recent Research in Freezing and Drying.* Blackwell Scientific Publications.

Prehoda, R. W. (1969). *Suspended Animation.* Chilton, Radnor, Pennsylvania.

Smith, Andrew (1961). *Biological Effects of Freezing and Super-*

cooling. (Monograph of the Physiological Society). Edward Arnold, London.

Smith, A. U. (Ed.) (1970). *Current Trends in Cryobiology*. International Cryogenics Monographs, Plenum Press, New York.

Solymar, L. (1972). *Superconductive Tunnelling and Applications*. Chapman and Hall, London.

Taylor, A. W. B. and Monkes, G. R. (1970). *Superconductivity*. Wykeham Publications, Taylor and Francis, London.

Williams, J. E. C. (1970). *Superconductivity and its Applications*. Pion Ltd., London.

Zemansky, Mark W. (1968). *Heat and Thermodynamics. An Intermediate Textbook*, 5th edition. McGraw Hill, New York.

Index